ENVIRONMENTAL SCIENCE, ENGINEERING AND TECHNOLOGY

WATER RESOURCES AND THE ARMY CORPS OF ENGINEERS: MANAGEMENT ACTIVITIES AND SAFETY PREPARATIONS

ENVIRONMENTAL SCIENCE, ENGINEERING AND TECHNOLOGY

Additional books in this series can be found on Nova's website under the Series tab.

Additional e-books in this series can be found on Nova's website under the e-book tab.

ENVIRONMENTAL SCIENCE, ENGINEERING AND TECHNOLOGY

WATER RESOURCES AND THE ARMY CORPS OF ENGINEERS: MANAGEMENT ACTIVITIES AND SAFETY PREPARATIONS

HARRY M. FOSTER
EDITOR

New York

Copyright © 2016 by Nova Science Publishers, Inc.

All rights reserved. No part of this book may be reproduced, stored in a retrieval system or transmitted in any form or by any means: electronic, electrostatic, magnetic, tape, mechanical photocopying, recording or otherwise without the written permission of the Publisher.

We have partnered with Copyright Clearance Center to make it easy for you to obtain permissions to reuse content from this publication. Simply navigate to this publication's page on Nova's website and locate the "Get Permission" button below the title description. This button is linked directly to the title's permission page on copyright.com. Alternatively, you can visit copyright.com and search by title, ISBN, or ISSN.

For further questions about using the service on copyright.com, please contact:
Copyright Clearance Center
Phone: +1-(978) 750-8400   Fax: +1-(978) 750-4470   E-mail: info@copyright.com.

NOTICE TO THE READER

The Publisher has taken reasonable care in the preparation of this book, but makes no expressed or implied warranty of any kind and assumes no responsibility for any errors or omissions. No liability is assumed for incidental or consequential damages in connection with or arising out of information contained in this book. The Publisher shall not be liable for any special, consequential, or exemplary damages resulting, in whole or in part, from the readers' use of, or reliance upon, this material. Any parts of this book based on government reports are so indicated and copyright is claimed for those parts to the extent applicable to compilations of such works.

Independent verification should be sought for any data, advice or recommendations contained in this book. In addition, no responsibility is assumed by the publisher for any injury and/or damage to persons or property arising from any methods, products, instructions, ideas or otherwise contained in this publication.

This publication is designed to provide accurate and authoritative information with regard to the subject matter covered herein. It is sold with the clear understanding that the Publisher is not engaged in rendering legal or any other professional services. If legal or any other expert assistance is required, the services of a competent person should be sought. FROM A DECLARATION OF PARTICIPANTS JOINTLY ADOPTED BY A COMMITTEE OF THE AMERICAN BAR ASSOCIATION AND A COMMITTEE OF PUBLISHERS.

Additional color graphics may be available in the e-book version of this book.

Library of Congress Cataloging-in-Publication Data

ISBN: 978-1-53610-420-2

Published by Nova Science Publishers, Inc. † New York

Contents

Preface		vii
Chapter 1	Army Corps of Engineers: Efforts to Assess the Impact of Extreme Weather Events *United States Government Accountability Office*	1
Chapter 2	Levee Safety: Army Corps and FEMA Have Made Little Progress in Carrying Out Required Activities *United States Government Accountability Office*	33
Chapter 3	Army Corps of Engineers: Actions Needed to Improve Cost Sharing for Dam Safety Repairs *United States Government Accountability Office*	51
Chapter 4	Army Corps of Engineers: Additional Steps Needed for Review and Revision of Water Control Manuals *United States Government Accountability Office*	79
Index		105

PREFACE

The U.S. Army Corps of Engineers (Corps) is the world's largest public engineering, design, and construction management agency, with water resources infrastructure in every state, authorized for various purposes including navigation, flood and coastal storm damage reduction, hydropower, and water supply. The Corps plans, designs, and operates water resources infrastructure projects, such as dams, levees, hurricane barriers, floodwalls, and floodgates, that may be affected by extreme weather events. This book discusses the Army Corps of Engineers and their water resource work in the areas of extreme weather events; levee safety; dam safety (cost sharing of repairs); and operations/water control manuals.

In: Water Resources ...
Editor: Harry M. Foster

ISBN: 978-1-53610-420-2
© 2016 Nova Science Publishers, Inc.

Chapter 1

ARMY CORPS OF ENGINEERS: EFFORTS TO ASSESS THE IMPACT OF EXTREME WEATHER EVENTS[*]

United States Government Accountability Office

ABBREVIATIONS

CEQ	Council on Environmental Quality
Corps	U.S. Army Corps of Engineers
CWMS	Corps Water Management System
NDBC	National Data Buoy Center
NEPA	National Environmental Policy Act of 1969
NOAA	National Oceanic and Atmospheric Administration
NRCS	Natural Resources Conservation Service
NWC	National Water Center
NWS	National Weather Service
SCAN	Soil Climate Analysis Network
SNOTEL	Snow Telemetry
USGS	U.S. Geological Survey
WRDA	Water Resources Development Act

---

[*] This is an edited, reformatted and augmented version of The United States Government Accountability Office publication, Report to Congressional Requesters GAO-15-660, dated July 2015.

WRRDA   Water Resources Reform and Development Act of 2014

Why GAO Did This Study

The Corps plans, designs, and constructs water resources infrastructure, such as dams and levees. According to the U.S. Global Change Research Program, the frequency and intensity of extreme weather events are increasing. Much of the Corps' infrastructure was built over 50 years ago and may not be designed to operate within current climate patterns, according to the U.S. Geological Survey.

The Water Resources Reform and Development Act of 2014 included a provision for GAO to study the Corps' management of water resources in preparation for extreme weather. This is the first in a series of reports GAO is issuing on this topic. GAO's other reports will examine operations and dam and levee safety, which GAO plans to issue in fiscal year 2016. This report explores (1) how the Corps prepares for and responds to extreme weather events in its planning and operation of water resources projects, and (2) the extent to which the Corps has assessed whether existing water resources infrastructure is prepared for extreme weather events. GAO reviewed Corps guidance on planning, operations, and assessments, and interviewed Corps officials from headquarters and eight districts— selected, in part, on number of projects.

What GAO Recommends

GAO previously recommended that the Corps work with Congress to develop a more stable funding approach. The Corps has not taken action, but GAO continues to believe the recommendation is valid. Agencies had no comments on a draft of this report.

What GAO Found

The U.S. Army Corps of Engineers (Corps) considers the potential impact of extreme weather events in its planning and operations of water resources infrastructure projects by, among other things, updating and developing guidance on how to incorporate different extreme weather scenarios in its planning of projects. For example, in 2014, the Corps issued guidance on how to evaluate the effects of projected future sea level change on its projects and what to consider when adapting projects to this projected change. In addition, Corps districts prepare water control manuals, guidance outlining project operations. The Corps can approve deviations from the manuals to alleviate critical situations, such as extreme weather events. For example, in December 2014, the Corps approved a deviation from operations at a southern California dam, which allowed the Corps to retain rainwater to help respond to the state's extreme drought conditions.

The Corps has assessed certain water resources infrastructure projects to determine whether they are designed to withstand extreme weather events. Specifically, the Corps has national programs in place to perform risk assessments on dams and levees, as required by law. Unlike the requirements for dams and levees, the Corps is not required to perform systematic, national risk assessments on other types of existing infrastructure, such as hurricane barriers and floodwalls and has not done so (see table). However, the Corps has been required to assess such infrastructure after an extreme weather event in response to statutory requirements, as it did in November 2013 and in January 2015, after Hurricane Sandy. Also, the Corps has performed initial vulnerability assessments for sea level rise on its coastal projects and has begun conducting such assessments at inland watersheds.

U.S. Army Corps of Engineers' Systematic, National Infrastructure Risk Assessments, 2006- June 2015

Type of infrastructure	Number of projects	Number of assessments
Dams	707	706
Levees (in segments)	2,887	1,232
Other[a]	[b]	0

Source: GAO analysis of Corps data.| GAO-15-660.
[a] Other includes infrastructure, such as hurricane barriers and floodwalls.
[b] The Corps has not yet completed an inventory of other types of infrastructure.

Unlike federal agencies that have budgets established for broad program activities, most Corps civil works funds are appropriated for specific projects. However, the Corps has not worked with Congress to develop a more stable funding approach, as GAO recommended in September 2010, which could facilitate conducting risk assessments. The Corps partially concurred with this recommendation, stating that it would promote efficient funding. As the frequency and intensity of some extreme weather events are increasing, without performing systematic, national risk assessments on other types of infrastructure, such as hurricane barriers and floodwalls, the Corps will continue to take a piecemeal approach to assessing risk on such infrastructure.

* * *

July 22, 2015

The Honorable James Inhofe
Chairman

The Honorable Barbara Boxer
Ranking Member
Committee on Environment and Public Works
United States Senate

The Honorable Bill Shuster
Chairman

The Honorable Peter DeFazio
Ranking Member
Committee on Transportation and Infrastructure
House of Representatives

The U.S. Army Corps of Engineers (Corps) is the world's largest public engineering, design, and construction management agency, with water resources infrastructure in every state, authorized for various purposes including navigation, flood and coastal storm damage reduction, hydropower, and water supply. The Corps plans, designs, and operates water resources infrastructure projects, such as dams, levees, hurricane barriers, floodwalls, and floodgates, that may be affected by extreme weather events. These extreme weather events include, among other things, flood, drought, and severe storms. According to the National Research Council and the U. S. Global Change Research Program's May 2014 National Climate Assessment, precipitation patterns are changing, and the frequency and intensity of some extreme weather events are increasing.[1] The Department of Commerce's National Oceanic and Atmospheric Administration (NOAA) found that, from 2010 through 2014, there have been 49 extreme weather events with losses exceeding $1 billion. Much of the Corps' infrastructure was built more than 50 years ago and, according to the Department of the Interior's U.S. Geological Survey (USGS), aging infrastructure may not meet its design level of performance under the current climate and could be more vulnerable to failure under future climate scenarios. Moreover, the USGS has reported that, of all the potential threats posed by changing weather patterns, those associated with water resources are arguably the most consequential for both society and the environment.[2] Federal agencies, such NOAA, USGS, and the Department of Agriculture's Natural Resources Conservation Service (NRCS), all collect and interpret weather and climate information the Corps uses in its management of water resources.

Section 3024 of the Water Resources Reform and Development Act of 2014 (WRRDA) included a provision for GAO to conduct a study of the strategies used by the Corps for management of water resources in preparation for and response to extreme weather. This report explores (1) how the Corps prepares for and responds to extreme weather events in its planning and operation of water resources infrastructure projects and (2) the extent to which the Corps has assessed whether existing water resources infrastructure is prepared for extreme weather events. This is the first in a series of reports we are issuing related to efforts by the Corps to manage water infrastructure resources. Specifically, we have work ongoing or commencing that will look more in depth at Corps project operations, dam safety, and levee safety, which we plan to issue in fiscal year 2016.

To determine how the Corps prepares for and responds to extreme weather events in its planning and operation of water resources infrastructure projects, we reviewed executive orders and Corps guidance on planning, operations, and assessments of infrastructure relating to extreme weather. For example, we reviewed Corps guidance for incorporating sea level rise into studies of new water resources projects, as well as water control manuals used to manage existing water resources projects. We reviewed and analyzed reports, studies, and plans from the Corps, including annual Corps Climate Change Adaptation Plans and the Corps' Climate Preparedness and Resilience Policy Statement. We interviewed Corps headquarters officials to obtain additional information about the agency's policies, procedures, and processes for planning, operations, and assessments of infrastructure to prepare for and respond to extreme weather events, as well as any challenges associated with incorporating that information. We also reviewed reports and spoke to agency officials that represent the Corps in interagency efforts such as the Climate Change and Water Resource Working Group.[3] In addition to officials from the Corps, we also interviewed officials from USGS, NOAA, and NOAA's National Weather Service (NWS), NRCS, and the Bureau of Reclamation to learn about how the Corps interacts with these agencies and what extreme weather data they obtain from these agencies.

To determine the extent to which the Corps has assessed whether existing water resources infrastructure is prepared for extreme weather events, we reviewed relevant Corps policies and guidance related to conducting infrastructure assessments. We also interviewed officials from the Corps' Response to Climate Change Program who led vulnerability assessments. We spoke to Corps officials responsible for carrying out the dam and levee safety programs to obtain additional information regarding the development of each program. Finally, we reviewed reports on the Corps' vulnerability assessments to learn the extent to which the Corps has considered extreme weather impacts on its existing water infrastructure projects.

In addition to conducting interviews with officials from Corps headquarters, for both of our objectives, we also interviewed officials from a nongeneralizable sample of eight districts to determine their procedures for incorporating information on extreme weather events into their planning, operations, and assessments.[4] We selected these district offices to get a range of perspectives based on geographical location, number of ongoing water resources infrastructure projects, as well as types of recent extreme weather events a district may have experienced from 2011 through 2014. We chose this time frame because implementing guidance for a 2009 executive order on climate change and internal Corps guidance on planning and operating infrastructure were released or updated during these years.

We conducted this performance audit from November 2014 to July 2015 in accordance with generally accepted government auditing standards. Those standards require that we plan and perform the audit to obtain sufficient, appropriate evidence to provide a reasonable basis for our findings and conclusions based on our audit objectives. We believe that the evidence obtained provides a reasonable basis for our findings and conclusions based on our audit objectives.

BACKGROUND

Located within the Department of Defense, the Corps has both military and civilian responsibilities.[5] Through its Civil Works program, the Corps plans, designs, constructs, operates, and maintains a wide range of water resources infrastructure projects for purposes such as flood control, navigation, and environmental restoration. The Civil Works program is organized into three tiers: a national headquarters in Washington, D.C.; eight regional divisions that were established generally according to watershed boundaries; and 38 districts nationwide (see fig. 1).

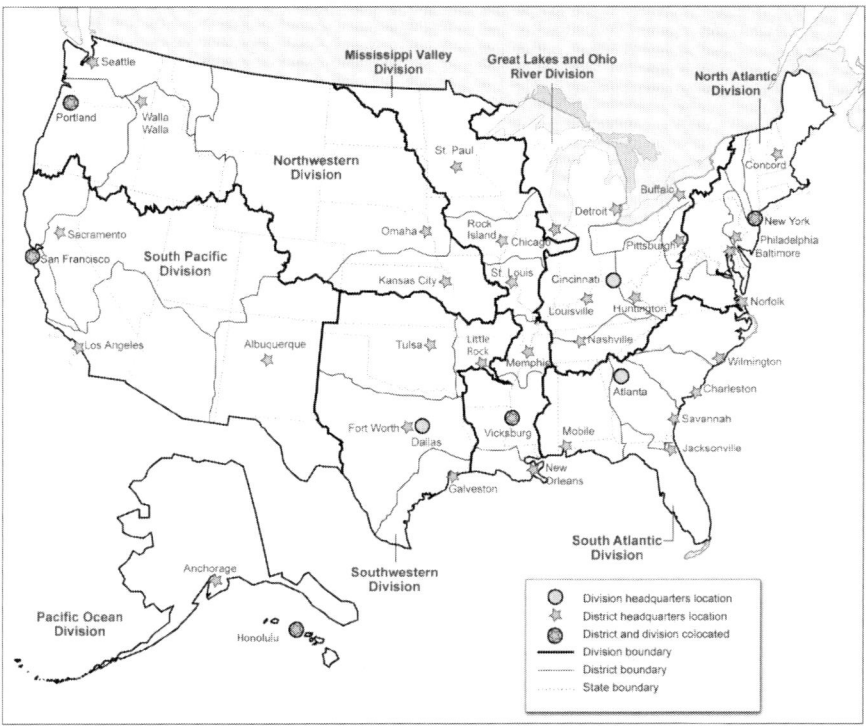

Source: U.S Corps of Army Engineers; Map Resources (map). | GAO-15-660.
Note: In addition to the eight divisions that manage U.S. water resources, the Corps operates a Europe District as well as a Transatlantic Division composed of the Middle East District and the Transatlantic Afghanistan District.

Figure 1. Locations of the U.S. Army Corps of Engineers' Civil Works Divisions and Districts.

Corps headquarters primarily develops policies and provides oversight. The Assistant Secretary of the Army for Civil Works, appointed by the President, establishes the policy direction for the Civil Works program.

The Chief of Engineers, a military officer, oversees the Corps' civil works operations and reports on civil works matters to the Assistant Secretary of the Army for Civil Works. The eight divisions, commanded by military officers, coordinate civil works projects in the districts within their respective geographic areas. Corps districts, also commanded by military officers, are responsible for planning, engineering, constructing, and managing water resources infrastructure projects in their districts. Districts are responsible for coordinating with the nonfederal sponsors, which may be state, tribal, county, or local governments or agencies. Each project has a project delivery team of civilian employees that manages the project over its life cycle. The team is led by a project manager and comprises members from the planning, engineering, construction, operations, and real estate functions.

In addition, the Civil Works program maintains a number of centers of expertise and research laboratories to assist the Corps divisions and districts in the planning, design, and technical review of civil works projects.[6] The Corps established these centers to consolidate expertise, improve consistency, reduce redundancy, and enhance institutional knowledge, among other things.

Unlike many other federal agencies that have budgets established for broad program activities, most Corps civil works funds are appropriated for specific projects. In general, the Corps receives "no-year" appropriations through the Energy and Water Development Appropriations Act—that is, there are no time limits on when the funds may be obligated or expended, and the funds remain available for their original purposes until expended. The conference report accompanying the annual Energy and Water Development Appropriations Act generally lists individual projects and specific allocations of funding for each project. Through this report, the appropriations committees essentially outline their priorities for the Corps' water resources projects. Congress directs funds for many individual projects in increments over the course of several years.

Steps for Developing Corps Projects

The Corps is responsible for planning, designing, and operating much of the nation's water resources infrastructure. To do so, the Corps generally goes through a series of steps involving internal and external stakeholders.

The Corps' Planning and Design Process

Usually, the Corps becomes involved in water resources infrastructure projects when a local community perceives a need or experiences a problem that is beyond its ability to solve and contacts the Corps for assistance. If the Corps does not have the statutory authority required for studying the problem, the Corps must obtain authorization from Congress before proceeding.[7] Studies have been authorized through legislation, typically a Water Resources Development Act (WRDA), or, in some circumstances, through a committee resolution by an authorizing committee. Next, the Corps must receive an appropriation to study the project, which it seeks through its annual budget request to Congress.

After receiving authorization and an appropriation, a feasibility study is to be conducted.[8] A Corps district office is to conduct a feasibility study, the cost of which is generally shared by a nonfederal sponsor, which may be a state, tribal, county, or local government, or agency. The feasibility study investigates the problem and makes recommendations on whether the project is worth pursuing and how the problem should be addressed. The district office is to conduct the study and any needed environmental studies and document the results in a feasibility report.

At specific points within the feasibility stage, a new infrastructure project is to undergo a series of technical reviews at the district, regional, and national level to assess the project's methodology and to ensure that all relevant data and construction techniques are considered. At the district level, all decision documents and their supporting analysis for a new project are to undergo a district quality control review by district leadership. This review is to assess the science and engineering work products to ensure that they are fulfilling project quality requirements. At the regional level, decision documents are to undergo an agency technical review by Corps officials from districts outside of the one conducting the study. This review verifies the district quality control review, assesses whether the analyses presented are technically correct and comply with published Corps guidance, and determines whether the documents explain the analyses and results in a reasonably clear manner for the public and decision makers. In some instances, a new project meeting certain criteria may also undergo an Independent External Peer Review.[9] For these Independent External Peer Reviews, the Corps is required by law to contract with the National Academy of Sciences, a similar independent scientific and technical advisory organization, or an "eligible organization" to establish a panel of experts that will review a project study.[10] Several criteria are used for selecting peer review panel members, including assessing and balancing members' knowledge, experience, and perspectives in terms of the subtleties and complexities of the particular scientific, technical, and other issues to be addressed.[11]

After going through various levels of review, depending on the project, the Chief of Engineers is to review the report and decide whether to sign a final decision document, known as the Chief's Report, recommending the project for construction. The Chief of Engineers is to transmit the Chief's Report and the supporting documentation to Congress through the Assistant Secretary of the Army for Civil Works and the Office of Management and Budget. Congress may authorize the project's construction in a WRDA or other legislation.[12]

Most infrastructure projects are authorized during the preconstruction engineering and design phase, which begins after the feasibility study is complete. The purpose of this phase is to complete any additional planning studies and all of the detailed technical studies and designs needed to begin construction of the infrastructure project.[13] Once the construction project has been authorized, the Corps seeks funds to construct the infrastructure project through the annual budget formulation process. As part of the budget process, the Army, with input and data from Corps headquarters, division, and district offices, develops a budget request for the agency. In fiscal year 2006, the Corps introduced what it refers to as performance-based budgeting, which uses performance metrics to evaluate projects' estimated future outcomes and gives priority to those it determines have the highest expected returns for the national economy and the environment, as well as those that reduce risk to human life. Congress directs funds for individual projects in increments over the course of several years. If the infrastructure project has been appropriated funds, the district enters into a cost-sharing agreement with the nonfederal sponsor.

The Corps' Operations and Maintenance

Once construction is completed, the Corps may turn over operation and maintenance of the infrastructure project to the nonfederal sponsor, which then bears the full cost, or the Corps may operate and maintain the project itself. Typically, the Corps operates and maintains reservoirs, locks, dams, and other water control infrastructure projects in which water storage is managed and operated for multiple purposes authorized by Congress such as flood control, navigation, recreation, hydropower, and other uses. For those projects, Corps guidance directs districts to develop a water control manual that is used to manage all of the projects' authorized purposes in consultation with interested stakeholders in the area of the project that may be impacted by its operations. In addition to water control manuals for individual projects, the Corps may also have master water control manuals that outline the operations of a system of projects.[14] The Corps may also develop operational guidance for non-Corps projects if, for example, the Corps has responsibility for flood control or other operations at that project. Water control manuals typically outline the operating criteria and guidelines for varying conditions and specifications for water storage and releases from a reservoir, including instructions for obtaining and reporting appropriate hydrologic data.

Weather-Related Data

The Corps uses a variety of hydrologic data—data relating to the movement and distribution of water—and forecasting data in its planning, designing, and operation of water resources infrastructure that can help it plan for extreme weather events. Much of these data are collected by other federal agencies as part of nationwide efforts to gather weather and hydrologic data. Table 1 shows examples of the type of hydrologic data collected by various federal agencies and used by the Corps.

Table 1. Selected Hydrologic Data Collected by Federal Agencies and Used by the U.S. Army Corps of Engineers to Manage Water Resources Infrastructure

Type of data	Agency	Program description	Types of data used by the Corps
Streamflow	United States Geological Survey (USGS)	The National Streamflow Information Program collects streamflow data through its national streamgage network[a] which continuously measures the level and flow of rivers and streams at 8,025 active continuous streamgages nationwide for distribution on the Internet.	Streamgages can provide information on streamflow as a discharge measurement (the amount of water moving through the river, for example, measured in cubic feetper second) or as a river stage measurement (the current height of the water in the river in feet) to monitor or make predictions.
Mountain snowpack	Natural Resources Conservation Service (NRCS)	NRCS operates 885 Snow Telemetry (SNOTEL) sites in 13 western states, which transmit snow depth, snow-water equivalent, and climate parameters in near real time. In addition, the NRCS snow course network conducts manual surveys of snow depth at	Both SNOTEL sites and snow courses gather information about snow depth, as well as the snow-water equivalent, which is the amount of water in the snowpack. This data is important to forecasting the water supply in the West.

Table 1. (Continued)

Type of data	Agency	Program description	Types of data used by the Corps
		about 956 sites in the United States.	
Plains snowpack	National Weather Service (NWS)[b]	The National Water Center (NWC) produces a map of snow conditions in the United States daily based on a combination of airborne surveys, satellite observations, and on-the-ground field measurements.[c]	NWC produces products, such as maps, of modeled snow-water equivalent across 31 states and 8 Canadian provinces. NWC also provides information about soil moisture based on their airborne surveys.
Precipitation	National Oceanic and Atmospheric Administration (NOAA)	NOAA collects snow and rain data in the continental United States with 143 weather radars and 10,000 precipitation gauges. Many of these gauges are owned and operated by other federal agencies; state, municipal, and tribal governments; and citizen observers.	Radar-detected precipitation and on-the-ground precipitation amounts from rain gauges are combined and analyzed to provide estimates of precipitation. Data collected by the U.S. Climate Reference Network is used in operationalclimate monitoring activities and for placing current climate anomalies into an historical perspective.
Soil moisture	NRCS	Through the Soil Climate Analysis Network (SCAN), NRCS collects information on climate and soil.	Soil moisture, in addition to soil temperature, air temperature, precipitation, wind speed, and direction, and many other climatic parameters, transmits data in near real time.
Oceanographic	NWS	Through the National Data Buoy Center (NDBC), NWS deploys buoys into the ocean to collect a variety of data.	Buoys collect data on wind direction and speed, wave height, and water temperature as well as other measures.

Source: GAO analysis of agency documents. | GAO-15-660.

[a] USGS's National Streamflow Information Program defines a streamgage as an active, continuously functioning device placed in a river or stream to measure water levels to aid in the estimation of mean daily streamflow throughout the year.

ᵇ In addition to federal efforts to collect plains snowpack data, some states have their own programs. For example, the North Dakota State Water Commission's Atmospheric Resource Board Cooperative Observer Network collects data on daily snowfall, monthly snowpack, and snow-water equivalent through a network of volunteer observers in North Dakota.

ᶜ The Corps manages a cooperative snow survey program that collects on-the-ground measurements of plains snowpack and snow-water equivalent. The Corps provides information from these snow surveys to NWC to help verify and validate the NWC models.

Corps officials and reports by federal agencies have highlighted limitations in some of the data the Corps uses in its planning, design, and operations of water resources infrastructure. Examples include the following:

- *Streamflow.* The Corps uses streamflow information from the National Streamflow Information Program in its planning, designing, and daily operations. However, according to Corps officials and USGS data, loss of streamgages due to funding constraints has reduced the available information about streamflows. According to USGS, from 1995 to 2008, 948 critical streamgages with 30 or more years of records were discontinued. Further, a USGS report noted that the loss of long-record streamgages reduces the potential value of streamflow information for infrastructure operations and design applications. Streamgage data are also used to produce climate change information upon which the Corps bases its adaptation planning. Despite these losses, the Corps has a formal agreement to provide funding to USGS to operate streamgages that provide data for the Corps' water management activities and in fiscal year 2013 provided USGS with $18 million.
- *Precipitation related to extreme storms.* Until 1999, the Corps used NWS Hydrometeorological Reports (site-specific probable maximum precipitation studies) for its designs. However, NWS discontinued providing these services in 1999 due to lack of funding, and some Corps officials said they have been using outdated data since that time. In response, the Corps has worked with the Interagency Federal Work Group on Extreme Storm Events since 2008 and established its own Extreme Storm Team to address Corps data needs, as well as the needs of other agencies.[15]
- *Plains snowpack.* The Corps uses plains snowpack data in its runoff forecasting for operations.[16] The Corps and NWS have found

limitations in this snowpack data. For example, a NWS report assessing the 2011 Missouri River flood found that modeled information on snow-water equivalent is available, but observational data are sparse and not always representative of basin-wide conditions.[17] WRRDA 2014 included a requirement that the Secretary of the Army, in coordination with other specified agencies, carry out snowpack and soil moisture monitoring in the Upper Missouri Basin.[18] As of June 2015, those agencies had not yet developed the monitoring system due to funding constraints according to agency officials.[19]

Executive Orders to Address a Changing Climate

Under Executive Orders 13514 and 13653, agencies are to create and update climate change adaptation plans that integrate consideration of climate change into their operations and overall mission objectives.[20] Specifically, Executive Order 13514, issued in 2009, directed agencies to participate in an existing Interagency Climate Change Adaptation Task Force. Based on the work of the task force, the Council on Environmental Quality (CEQ) issued implementing instructions for the executive order in March 2011.[21] The instructions directed agencies to, among other things, issue an agency-wide climate change adaptation policy statement and submit their climate adaptation plans to CEQ and the Office of Management and Budget. Executive Order 13653, issued in 2013, directed agencies to continue developing and regularly updating their climate adaptation plans.

In response to these executive orders, the Corps submitted its Climate Change Adaptation Plans in 2012, 2013, and 2014 (adaptation plan). The Corps' adaptation plan is implemented, in part, through its Responses to Climate Change Program. This program is charged with developing the methods, tools, and guidance to improve the resilience of the Corps' built and natural infrastructure through a collaborative, proactive, nationally consistent, and regionally sensitive framework and program of actions. According to the adaptation plan, these actions include improving the agency's understanding of climate impacts to missions and operations, assessing vulnerabilities, and identifying specific actions to minimize risk and capitalize on opportunities to improve infrastructure resilience. According to Corps documents, infrastructure resilience is the ability to anticipate, prepare for, respond to, and adapt to changing conditions and to withstand and recover rapidly from disruptions with minimal damage. As directed by CEQ instructions and guidance implementing Executive Order 13514, the Assistant Secretary of the Army for Civil Works released the Corps' policy regarding adaptation in June 2014.[22] The policy states that "mainstreaming climate change adaptation means that it will be considered at every step in the project life cycle for all USACE [U.S. Army Corps of Engineers] projects, both existing and planned ... to reduce vulnerabilities and enhance the resilience of our water resource infrastructure." This policy also established the Corps' Committee on Climate Preparedness and Resilience to oversee and coordinate the agency's climate change adaptation planning and implementation.

In January 2015, Executive Order 13690 was issued establishing a federal flood risk management standard, which applies to federal actions—including the construction of facilities with federal funds—in, and affecting, floodplains. Under the standard, certain new construction, substantially improved structures, and substantially damaged projects must meet a certain elevation level, among other things. Draft floodplain management guidelines were issued in February 2015 and were available for public comment through May 6, 2015. Within 30 days of the close of this public comment period, Executive Order 13690 directed agencies to submit an implementation plan to the National Security Council that contains milestones and a timeline for implementation of the executive order and standard.[23] However, federal agencies are prohibited from using appropriated funds to implement the standard until input from Governors, Mayors, and other stakeholders has been solicited and considered. According to the executive order, agencies should not issue or amend regulations and procedures to implement the executive order until after implementing guidelines are issued. Thus, it is unclear how the standard will affect the Corps' operations.

THE CORPS CONSIDERS THE POTENTIAL IMPACT OF EXTREME WEATHER IN ITS PLANNING AND OPERATIONS OF WATER RESOURCES INFRASTRUCTURE PROJECTS

The Corps addresses the potential impact of extreme weather events in its planning and operations of water resources infrastructure projects in various ways including updating and developing guidance to be used in the planning process; using tools, such as water control manuals, in its operation of projects; and through collaboration with key federal agencies and stakeholders.

Corps Guidance and Initiatives Call for the Consideration of the Potential Impacts of Extreme Weather Events in the Planning Process

The Corps considers the potential impacts of extreme weather in its planning process by updating and developing guidance, as well as incorporating the uncertainties of extreme weather events in planning for new infrastructure projects, and through its Civil Works Transformation Initiative. For example, in 2009, the Corps issued guidance for incorporating sea level change in its planning, construction, and operation of water resources infrastructure projects impacted by the rise and fall of sea levels.[24] This guidance, which was updated in 2011 and 2013, directs Corps districts to consider three scenarios of potential sea level change when designing and constructing new infrastructure, as well as managing existing water infrastructure.[25] According to Corps documents, sea level change can have a number of impacts on coastal and estuarine zones, including more severe storm and flood damages.[26] In 2014, the Corps issued additional guidance on how to evaluate the effects of projected future sea level change on Corps projects and what to consider when adapting projects to this projected change.[27] This guidance is intended to incorporate sea level change into the planning process to improve the resilience of projects and maximize performance over time.

In addition, in May 2014, the Corps issued guidance for how to incorporate potential impacts of extreme weather into the planning of inland infrastructure projects in accordance with Executive Order 13653 and the President's Climate Action Plan.[28] This guidance outlines the purpose and objective for incorporating this consideration into current and future studies as well as provides an example of how to incorporate new science and engineering in hydrologic analyses for new and existing Corps projects. Moreover, the guidance establishes a procedure to perform a qualitative analysis of potential climate threats and impacts to the Corps' hydrology-related projects and operations. The guidance calls for districts to conduct an initial screening-level qualitative analysis to identify whether climate change is relevant to the project goals or design. If climate change is determined to be relevant to the project goals or design, the guidance directs districts to make an evaluation of information about climate change impacts such as changes in processes governing rainfall runoff or snowmelt. The information is intended to be used to help identify opportunities to reduce potential vulnerabilities and increase resilience as a part of the project's authorized operations, as well as identify any limitations or issues associated with the data collected.

The Corps also issued guidance in October 2014 on determining the appropriate use of paleoflood information in its planning and operation of water infrastructure. [29] According to Corps guidance, useful information can be gained from paleohydrology, or the evidence of the movement of water and sediment in stream channels before continuous hydrologic records or direct measurements became available. For example, this information can be derived from high water marks, tree rings, and gravel deposits, among other things, and can help Corps districts estimate flood peak magnitudes, volumes and durations for flood damage assessments, or evaluate design criteria. This guidance also notes that paleoflood information may not be suitable for all projects such as, watersheds that have been altered through time, either by geologic processes or by human activity.

In addition to updating and developing guidance for planning and operating water infrastructure, Corps headquarters officials told us that they also have taken steps to incorporate uncertainty, such as that associated with extreme weather, into their planning process through the Civil Works Transformation Initiative. According to Corps documents, the Civil Works Transformation Initiative began in 2012 to aid the Corps in meeting current and future challenges and addressing the water resources needs throughout the United States. As part of the Initiative, the Corps updated its planning process in 2012 to help strengthen the incorporation of risk into planning assumptions for feasibility studies on new infrastructure projects. For example, Corps headquarters officials told us that they have adopted a risk-informed approach to help address uncertainty, such as that associated with extreme weather, by defining the levels of risk associated with a variety of project designs. Corps officials said, beginning in 2012, feasibility studies for new projects have used this approach to identify risks, including extreme weather, which may occur throughout the life cycle of a water resources project. Headquarters officials told us that, under this approach, project delivery teams must address risks associated with climate change in their project planning documents. To help ensure that the appropriate weather and climate data are being used in the planning process, since 2012, the Corps' external peer review process has asked experts to review the project plans and note whether appropriate data and information were used to respond to extreme weather risks. Corps officials told us that independent external review questions relating to climate change differ, depending on when they were prepared, as well as the type of input provided by the project delivery team and district officials. Because the Civil Works Transformation Initiative is not yet complete, it may be too early to evaluate the impact of this initiative.

The Corps Uses Water Control Manuals and Other Tools to Help Prepare for Extreme Weather

The Corps uses a variety of tools in its operations to help prepare for extreme weather, including water control manuals and an automated information system. Water control manuals, which outline the operation of water storage at individual projects, or a system of projects, are used by the Corps to prepare for extreme weather events. These manuals are to outline the various types of weather-related data the Corps uses in its daily operations, as well as when extreme weather events occur. The manuals are also to describe the automated processes used in a data exchange with USGS and the regional NWS center that provides weather forecasts, rainfall information, and streamflow data, among other information to the Corps to prepare for extreme weather events. In addition, water control manuals include a description of the historical information that is used for purposes of creating models to predict streamflow and reservoir stages. Corps guidance, in the form of engineer regulations, describes what is to be included in water control manuals, such as directing districts to establish and outline special operational practices during emergency situations, as well as a drought contingency plan. According to Corps officials, this Corps guidance ensures that Corps districts' water control manuals are created in a standardized manner so all districts are prepared for extreme weather events. Corps guidance also directs districts to ensure that all authorized purposes of a project are addressed in its operations and notes that operations must strike a balance among those purposes, which often have competing needs. According to the Corps' engineer regulations, any operational priorities among multiple authorized purposes during extreme conditions, such as drought or flooding events, may need to be defined in water control manuals.

According to the Corps guidance, water control manuals also must contain provisions for the Corps to temporarily deviate from operations, when necessary, to alleviate critical situations. According to Corps officials, critical situations may include extreme weather events, such as a flood or drought. For example, in December 2014, the Corps approved a deviation from operations at Prado Dam in southern California, which allowed the Corps to temporarily retain water captured behind the dam following a rainstorm. This deviation, along with other deviations in the southern California region, was in response to the drought that California has experienced since 2011. According to Corps guidance, deviations are meant to be temporary and, if a deviation lasts longer than 3 years, the water control manual must be updated. Corps officials we spoke with were unaware of any deviations that, as of May 2015, have lasted more than 3 years.

Corps headquarters and district officials we interviewed said that some water control manuals may need to be updated due to changing conditions in the watershed; however, they also said that some manuals in existence for many years may not necessarily need to be updated since, in part, they allow for flexibility with changing weather trends. Specifically, headquarters and district officials we spoke with said projects that have not experienced a change in land use around the basin, a change in climate patterns, or new weather-related information may not need to be revised. Furthermore, headquarters officials said water control manuals, including reservoir rule curves and drought contingency plans, have proved relatively robust to the climate changes already observed in the West.[30] According to these officials, when combined with the ability to temporarily deviate from operations, when necessary, there is flexibility to respond to short-term and long-term needs based on the best available information and science. The Corps is currently working to develop and implement a strategy to update drought contingency plans to account for climate change. According to Corps officials, the agency will complete its strategy for updating these plans by fiscal year 2016.

Corps guidance directs districts to periodically review and revise water control manuals, as necessary, to conform to changing requirements resulting from land development in the project area, improvements in technology, and the availability of new hydrologic data, among other things. Some district officials said water control manuals have not been consistently updated due to changing conditions in the watershed, primarily due to funding constraints. Corps headquarters officials said there is not a Corps-wide process in place to assess whether manuals should be updated; rather, it is up to the discretion of the districts to do so. Some district officials said that they had requested funding to update water control manuals but did not receive the requested funding to conduct such updates. We will continue to assess this issue.[31]

The Corps has also established the Corps Water Management System (CWMS), an automated information system supporting the Corps' operations to, among other things, prepare for extreme weather. CWMS contains various data, such as weather conditions, soil moisture, snow accumulation, streamflow, and water level that can be used by the districts to develop models of watershed and channel processes and to forecast future availability of water. For example, CWMS allows the districts to simulate different operational scenarios to determine which one will more likely result in higher downstream water levels due to a large storm. According to Corps documentation, information from the simulation is intended to help the districts assess the economic, environmental, life safety, and other consequences, such as those from an extreme weather event, of different operational scenarios and lead to better-informed operational decisions. For example, Los Angeles district officials told us that CWMS models are being calibrated for expected maximum flood conditions which can allow them to better forecast runoff volumes in areas prone to extreme weather events. According to Corps documents, CWMS also will support rapid flood forecasting by the district and help reduce the potential for flooding in the basin. CWMS has been deployed to 35 of 38 districts since 2009 but has not yet been fully integrated into all Corps districts, and the watershed and channel models have not been fully implemented as of June 2015. The Corps plans to complete integrating CWMS into all districts by the end of 2015, and an effort is under way to have the watershed and channel models fully integrated by 2023 or earlier, depending on funding.

The Corps Collaborates with Key Stakeholders to Plan and Operate Water Infrastructure

The Corps has taken steps to prepare for extreme weather through its participation in various collaboration efforts with federal agencies and other stakeholders at both the regional and national levels.

At the regional level, Corps district officials told us that their collaboration with federal agencies and local stakeholders is sufficient for effective planning and operation of water infrastructure. Corps officials told us they regularly collaborate with federal agencies and local stakeholders to help ensure that they have the weather and climate data needed to plan and operate water infrastructure and to address extreme weather in a coordinated manner. For example, Alaska district officials told us their district has a long history of collaborating with NWS and USGS to monitor data across the remote areas of Alaska and now collaborates with these agencies using a geostationary satellite.

Little Rock district officials told us they participate with agency officials from USGS and NOAA, as well as other stakeholders, at Tri-Agency Fusion Team meetings to discuss ways to improve the accuracy of the data generated by the agencies and improve the accuracy and utility of rainfall observations and river forecasts. Savannah district officials told us they regularly communicate with NWS officials in advance of and during extreme weather events. Within certain regions, Corps district officials told us they regularly interact with state and federal officials through the Silver Jackets program to, among other things, identify gaps among agency programs, leverage information and resources, and provide access to national programs such as the Corps Levee Inventory and Assessment Initiative.[32] The Corps is also conducting regional pilot studies nationwide to test different methods and frameworks for adapting to climate change in which they involve numerous stakeholders. For example, four Corps districts completed an Ohio River Basin pilot study in 2013 in which the districts worked with more than 70 stakeholders, including federal and state agencies, academia, and private entities. The pilot study considered the potential effects of climate change on future management of water resources, including 83 Corps dams, 131 levees and floodwalls, and 63 navigational locks in the 204,000 square miles of the basin. As a result of this pilot study, a consortium of basin interests convened the Ohio River Basin Alliance to address common interests in water resources and basin-wide climate change issues.

Corps district officials told us that they may also interact with state agencies, universities, and private industry to collect data that may not be collected by federal agencies. For example, Little Rock district officials told us they have used the Community Collaborative Rain, Hail, and Snow Network, in which precipitation data are collected by volunteer citizens and published daily on the Internet by Colorado State University. Walla Walla district officials have been participating with the University of Washington since 2012 in support of the Columbia River Treaty analysis that involves information on data collection, modeling, and trends on future weather and climate changes predicted for the region. Some districts told us they also have gained valuable and up-to-date technical information on engineering and design techniques from private industry associations and made key contacts at industry conferences. However, all the districts we spoke with told us they face challenges in attending weather-related conferences sponsored by entities other than the federal government due to changes in Department of Defense conference policies.[33]

Corps officials also collaborate with other federal agencies and stakeholders at the national level to identify data gaps that may exist and disseminate critical water resource information and data. For example, Corps headquarters officials have participated in the Climate Change and Water Working Group, a working-level forum established to share information and accelerate the application of climate information in water management, among other things. Through this group, the Corps along with local, state, and federal water management agencies, have examined water user needs for climate and weather information for long- and short-term water resources planning and management and have issued two reports on their findings.[34] We have previously reported the Corps along with NOAA, USGS, and other stakeholders developed the Federal Support Toolbox, a federal Internet portal, to provide current, relevant, and high-quality information on water resources and climate change data applications and tools for assessing the vulnerability of water programs and facilities to climate change.[35] The toolbox is publicly available online through the Integrated Water Resource Science and Services group and is maintained by the Corps with contributions from more than 16 federal agencies and nongovernment partners. According to agency officials, the Integrated Water Resource Science and Services group consists of four core agencies (USACE, NWS, USGS, and the Federal Emergency Management Agency) and is currently focused on improvements of water forecasting and integration of related models and databases.

THE CORPS HAS ASSESSED CERTAIN INFRASTRUCTURE PROJECTS TO PREPARE FOR EXTREME WEATHER EVENTS

The Corps has assessed certain water resources infrastructure projects to determine whether they are designed to withstand extreme weather events. Specifically, the Corps has national programs in place to perform risk assessments on dams and levees, as required by law, but, unlike these programs, the Corps is not required to perform systematic, national risk assessments on other types of infrastructure, such as floodwalls and hurricane barriers and has not done so. However, Corps officials said they have been required to assess such infrastructure after an extreme weather event in response to statutory requirements. The Corps has also performed some preliminary vulnerability assessments for sea level rise on its coastal projects and is beginning to conduct vulnerability assessments of inland watersheds to determine how a changing climate is affecting those projects.

The Corps Performs Risk Assessments on Dams and Levees but Does Not Assess Other Types of Infrastructure

The Corps performs risk assessments of its dams and levees through two national programs—the Dam Safety Program and the Levee Safety Program—but does not have similar programs in place for other types of infrastructure.[36] As part of its Dam Safety Program, from 2005 to 2009, the Corps performed a screening of 706 of its 707 dams to determine which of its five risk classifications those dams fell under—very high urgency, high urgency, moderate urgency, low urgency, and normal.[37] This risk classification addresses the probability of failure and resulting potential consequences due to failure. Part of the assessment determines whether the dams are designed and operated in such a way that, during a potential flood event, the downstream flooding would not be more severe than flooding if the dam did not exist. The risk assessment also takes into account the likelihood of an extreme weather event. According to Corps officials, all Corps-operated dams will undergo periodic assessments every 10 years because the risk at any given dam may change over time. The Corps has also established the Risk Management Center as a resource to manage and assess risks to dams and levee systems, and the Dam Safety Modification Mandatory Center of Expertise to provide technical advice, oversight, review, and production capability to districts performing any dam modifications in response to the risk assessment. The Dam Safety Modification Mandatory Center of Expertise also maintains a list of subject matter experts in the field of dam safety whose names are accessible via the Internet.

The Corps assesses the risk of its dams through the Dam Safety Program, but not all dam safety modification projects have been funded. More specifically, the Corps' initial screening of dams, completed in 2009, found that 18 dams fell under the very high urgency classification, 83 dams fell under the high urgency classification, 219 dams fell under the moderate urgency classification, and the remaining 386 dams fell under the low urgency classification. For those dams in the very high urgency, high urgency, and moderate urgency classifications, the Corps guidance directs that an Interim Risk Reduction Measures Plan be developed, which is a temporary approach to reduce dam safety risks while long-term solutions are being pursued. The

Corps found that completing dam safety modifications on its dams in the three most urgent classifications would cost more than $23 billion.[38] The cost for dam safety modifications for the very high urgency classification was about $4.2 billion; the high urgency classification cost was about $7 billion, and the moderate urgency classification cost was about $12 billion. According to Corps officials, from fiscal year 2009 through fiscal year 2014, the Corps received about $2.5 billion in appropriations to begin dam safety modification studies and construction on 15 very high urgency dams. As of June 2015, dam safety modification construction has been completed on seven very high urgency dams, and the Corps was working on the other eight.[39] Corps officials we spoke with in two districts said that they recognize that dams in other districts may fall into the very high urgency classification for modifications, but the dams in their own district, which are in the high urgency classification, are also at a high risk of failure should an extreme weather event occur.

The Corps also operates the Levee Safety Program, which began in 2006. According to the Corps, although the Dam and Levee Safety Programs are similar in their approach to risk assessments, the Levee Safety Program has not progressed as quickly, largely because the Corps owns and operates less than 20 percent of the 14,700 miles of levees that fall under the program.[40] Until 2009, the Corps collected information on 14,700 miles of levees for inclusion in the National Levee Database.[41] Since that time, the Corps has been conducting risk assessments of the 14,700 miles of levees that are included in the Levee Safety Program. The 14,700 miles of levees are divided into 2,887 segments, and risk assessments have been completed for about 43 percent of those segments as of April 2015. The Levee Safety Program risk assessments are to take into account the likelihood of an extreme weather event and how a levee will perform during that event. Based on the risk assessments that have been completed as of April 2015, 1 percent of those levees are classified as very high urgency, 8 percent are classified as high urgency, 27 percent are classified as moderate urgency, and 64 percent are classified as low urgency. Based on the risk assessments completed, as of June 2015, the Corps has not begun making improvements to any of the levees it owns and operates because it is still conducting the risk assessments and will prioritize any improvements once those assessments are complete. Improvements made to the non-Corps levees based on the results of the risk assessment are at the discretion of the local sponsor, with advice from the Corps on risk reduction measures.

Unlike the requirements for the Dam Safety and Levee Safety Programs, the agency is not required to perform risk assessments on other types of existing infrastructure, such as hurricane barriers and floodwalls, and it has not yet conducted an inventory of other types of infrastructure.[42] According to Corps officials, the agency has not performed systematic, national risk assessments on other types of existing infrastructure given funding limitations (see table 2). However, the Corps has received appropriations for and has been required to assess such infrastructure after an extreme weather event, such as in the aftermath of Hurricane Katrina in 2005 and Hurricane Sandy in 2012.[43]

Table 2. U.S. Army Corps of Engineers' Systematic, National Infrastructure Risk Assessments, 2006-June 2015

Type of infrastructure	Number of projects	Number of assessments
Dams	707	706
Levees (in segments)	2,887	1,232
Other[a]	[b]	0

Source: GAO analysis of Corps data. | GAO-15-660.
[a] Other includes infrastructure, such as hurricane barriers and floodwalls.
[b] The Corps has not yet completed an inventory of other types of infrastructure.

Subsequent to Hurricane Sandy, for example, the Corps released, in November 2013, an assessment of the performance of specific projects and, in January 2015, a more general assessment of the North Atlantic coastline. The project-specific performance assessment evaluated 75 constructed coastal storm risk management projects in the Corps' North Atlantic Division, which extends from Maine to Virginia, 31 projects in the Great Lakes and Ohio River Division, and 9 projects in the South Atlantic Division. For the more general assessment, the Corps looked at the risk along 31,000 miles of Atlantic Ocean shoreline from Virginia to New Hampshire as a system.[44] The Corps divided the area into multiple areas of coastline that were hydraulically separate from one another, studying the risk of flood, as well as the exposure of the populations, exposure by population density, infrastructure density, vulnerability by socioeconomic factors, and vulnerability of environmental resources and cultural resources. This risk assessment identified, among other things, nine high-risk areas of the North Atlantic Coast that warrant additional analyses to address coastal flood risk. As of June 2015, the Corps has made no improvements to its projects based on the general risk assessment in the Corps' study of Hurricane Sandy, which was made final in January 2015. However, according to Corps officials, many projects identified in the project-specific assessment received funding for and received repair and restoration through the Corps' Flood Control and Coastal Emergencies Program.[45]

The Corps conducted these risk assessments following Hurricanes Katrina and Sandy after receiving an appropriation for and being required by law to conduct them, as the Corps does not generally receive funding for broad program activities, such as risk assessments on infrastructure other than dams and levees. However, the Corps has not worked with Congress to develop a more stable funding approach, as we recommended in September 2010, which could facilitate such risk assessments.[46] That report found that a more stable funding approach could improve the overall efficiency and effectiveness of the Civil Works program. The department partially concurred with our recommendation, stating that it would promote efficient funding. As the frequency and intensity of some extreme weather events are increasing, without performing risk assessments on other types of existing infrastructure, such as hurricane barriers and floodwalls, before an extreme weather event (e.g., using a risk-based model), the Corps will continue to take a piecemeal approach to assessing risk on such infrastructure.[47] For this reason, we continue to believe our recommendation is valid.

The Corps Has Conducted Some Vulnerability Assessments of Water Infrastructure Projects to Prepare for Extreme Weather Events

The Corps has conducted two nationwide screening level assessments to assess its vulnerability to climate change in its management and operation of water infrastructure. According to the Corps' 2014 Climate Adaptation Plan, these vulnerability assessments are necessary so the Corps can address a changing climate and successfully perform its missions, operations, programs, and projects in an increasingly dynamic environment.

In 2013, the Corps began an initial project-level vulnerability assessment for coastal projects relating specifically to sea level change. Teams from 21 Corps districts with coastal projects reviewed more than 1,431 projects to determine the impact of sea level change at the 50- and 100-year planning horizons for coastal projects.[48] These projects were given a score based on science-based parameters to categorize the level of impact that sea level change would have on each project. The Corps completed these initial vulnerability assessments for coastal projects in September 2014 and determined that 944 of the 1,431 projects appear to be able to withstand future changes resulting from sea level rise, 94 projects may experience high or very high impacts as a result of sea level rise, and 393 projects may experience a low or medium impact as a result of sea level rise. As of June 2015, the Corps had begun prioritizing the 94 projects that may experience high or very high impacts as a result of sea level rise for a more detailed assessment. Corps officials said they do not yet know when this prioritization will be completed.

As of June 2015, the Corps was piloting methods to conduct the more detailed vulnerability assessment. In one pilot, through a vulnerability assessment of a hurricane barrier in New England that was designed in 1962 to provide navigation and flood risk reduction benefits for the area surrounding a harbor, the Corps found the project had experienced a 6- inch loss from its design elevation due to sea level rise. The hurricane barrier was listed as having potentially high impact from sea level rise in the screening assessment. The more detailed pilot assessment identified a potential future loss of elevation of between 6 inches and 2 feet 3 inches by 2065. Based on Corps data, the change in sea level has resulted in a reduction in the distance between the top of the water and the top of the hurricane barrier from 17 feet at its design to 16.5 feet currently, and potentially down to 14.25 feet within 50 years (see fig. 2). According to Corps officials, these future changes in distance between the top of the sea and the top of the hurricane barrier can result in a greater risk of floods and more operations of the navigation gate, which in turn reduces navigation reliability and increases maintenance costs. The Corps had initially planned to release a draft report on the initial coastal vulnerability assessment in December 2014 but, as of June 2015, the final report had not been released. Corps officials said the final report will likely be released in late summer 2015.

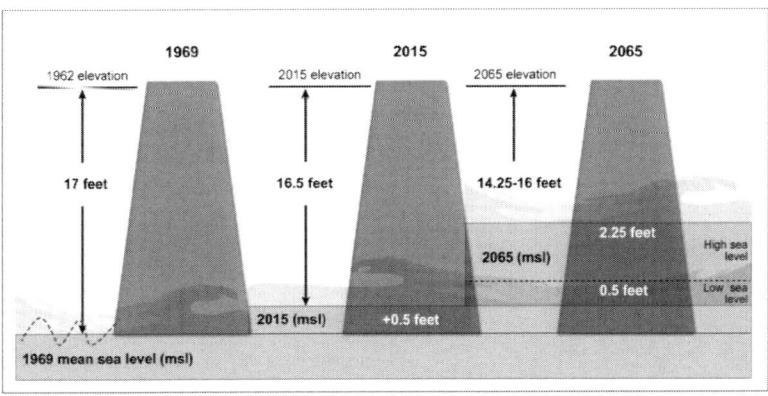

Source: GAO representation of U.S. Army Corps of engineers data. | GAO-15-660.

Figure 2. U.S. Army Corps of Engineers' Hurricane Barrier Design Elevation Changes Due to Sea Level Rise.

Corps officials acknowledge that the science is not yet available to conduct project-level vulnerability assessments for inland projects. However, the Corps initiated a study in 2012 that focused on how hydrologic changes due to climate change may impact freshwater runoff in some watersheds. As of June 2014, the Corps had identified the top 20 percent of watersheds that were most vulnerable for each business line through this initial watershed study.49 Corps officials said this is an initial screening level assessment that will lead to more detailed assessments of the most vulnerable water resources infrastructure projects and those with the highest potential impact from extreme weather events. The Corps is working with an expert consortium of federal, academic, nongovernmental organizations, and others to develop the climate and hydrology information necessary to conduct project-level assessments. Corps officials said that the consortium will develop the information needed to perform the project assessments and that it is unclear how long developing the necessary science will take.

According to the Corps' 2014 Climate Change Adaptation Plan and Corps headquarters officials, the inland and coastal vulnerability assessments will be merged over the next several years and will be used to determine how the Corps needs to manage and plan for new water resource projects. As of May 2015, Corps officials told us that they did not have a timeline for merging these assessments, in part because the climate and hydrology information is not yet available.

AGENCY COMMENTS AND OUR EVALUATION

We provided a draft of this report for review and comment to the Departments of Agriculture, Commerce, Defense, and the Interior for comment. These agencies did not provide written comments. In an e-mail received on June 29, 2015, the audit liaison for NOAA at the Department of Commerce provided technical comments for our consideration. In addition, in oral comments received on July 2, 2015, the Corps' point of contact on the engagement provided technical comments for our consideration. We incorporated these technical comments as appropriate.

Anne-Marie Fennell
Director,
Natural Resources and Environment

End Notes

[1] Jerry M. Melillo, Terese (T.C.) Richmond, and Gary W. Yohe, eds., 2014: Climate Change Impacts in the United States: The Third National Climate Assessment. U.S. Global Change Research Program (Washington D.C.: May 2014).

[2] Harry F. Lins, Hirsch, Robert M., and Kiang, Julie, Water—the Nation's Fundamental Climate Issue: A White Paper on the U.S. Geological Survey Role and Capabilities: U.S. Geological Survey Circular 1347 (2010).

[3] According to its website, the Climate Change and Water Resource Working Group is a working-level forum that fosters communication and research partnerships around these needs across the water and science communities of practice. The participating federal agencies include: the Bureau of Reclamation, the Corps, the National Oceanic and Atmospheric Administration, the U.S. Geological Survey, the Environmental Protection Agency, the Federal Emergency Management Agency, the National Aeronautics and Space Administration, and the U.S. Department of Agriculture.

[4] We spoke with Corps officials from the following districts: Baltimore, Maryland (North Atlantic Division); Savannah, Georgia (South Atlantic Division); Vicksburg, Mississippi (Mississippi Valley Division); Louisville, Kentucky (Great Lakes and Ohio River Division); Little Rock, Arkansas (Southwestern Division); Walla Walla, Washington (Northwestern Division); Los Angeles, California (South Pacific Division); and Anchorage, Alaska (Pacific Ocean Division).

[5] The Corps' Military program provides, among other things, engineering and construction services to other U.S. government agencies and foreign governments. This report only discusses the Civil Works program.

[6] The Centers of Expertise program is two-tiered with centers being mandatory or voluntary. For a full list of the Corps' centers of expertise, see http://www.usace.army.mil/about/centersofexpertise.aspx.

[7] If the Corps has previously performed an evaluation in the geographic area for a similar purpose, a new study can be authorized by an authorizing committee resolution. If the Corps has not previously investigated the area, the study needs to be authorized through legislation.

[8] Pub. L. No. 113-121, § 1002(a)(2), 128 Stat. 1193 (2014). Prior to WRRDA 2014, the Corps conducted studies in two phases: reconnaissance and feasibility. The reconnaissance study was conducted at full federal expense to determine if the problem warranted federal participation in a feasibility study and how the problem could be addressed. WRRDA 2014 eliminated the reconnaissance phase to accelerate the study process and allow the Corps to proceed directly to the feasibility study.

[9] Section 2034 of WRDA 2007 as amended (Pub. L. No. 110-114, § 2034 (2007), as amended by Pub. L. No. 113-121, § 1044 (2014)). A project must be subject to Independent External Peer Review if it meets at least one of the following criteria, and it is not a project excluded from peer review: the project has an estimated total cost of more than $200 million; the Governor of an affected state requests an independent peer review; or the Chief of Engineers determines that the project study is controversial (i.e., significant public dispute exists as to the project's size, nature, or effects, or its economic or environmental costs or benefits). In addition, the Chief of Engineers must consider a project for peer review if heads of certain federal or state agencies request it.

[10] Section 2034(I)(3) of WRDA 2007 defines an eligible organization as one that: (1) is a 501(c)(3) tax-exempt organization, (2) is independent, (3) is free from conflicts of interest,

(4) does not carry out or advocate for or against federal water resources projects, and (5) has experience in establishing and administering peer review panels.

[11] GAO examined the Corps' project peer review process in GAO, Army Corps of Engineers: Peer Review Process for Civil Works Project Studies Can Be Improved, GAO-12-352 (Washington, D.C.: Mar. 8, 2012).

[12] Corps officials estimated that about 75 percent of all feasibility studies result in a project authorized for construction.

[13] Section 2035 of WRDA 2007 as amended (Pub. L. No. 110-114, § 2035 (2007), as amended by Pub. L. No. 113-121, § 3028 (2014)) requires the Chief of Engineers to ensure that the design and construction activities for hurricane and storm damage reduction and flood damage reduction projects have a safety assurance review by independent experts if the Chief of Engineers determines that such a review is necessary to assure public health, safety, and welfare, prior to initiation of physical construction and periodically thereafter until construction activities are completed.

[14] There may be multiple projects located along a river and several tributaries. In such cases, a master water control manual is prepared to define system regulation.

[15] For example, the Extreme Storm Team is working to develop up-to-date hydrometeorological reports, create regional and site-specific probable maximum precipitation studies, and create an extreme storm catalog as a data archive of extreme storm events.

[16] Runoff flows over the land surface, going downhill into rivers and streams.

[17] National Oceanic and Atmospheric Administration, National Weather Service, Service Assessment: The Missouri/Souris River Floods of May – August 2011 (May 2012).

[18] Pub. L. No. 113-121, § 4003(a)(1)(A), 128 Stat. 1193, 1311 (2014).

[19] GAO, Missouri River Basin: Agencies' Progress Improving Water Monitoring Is Limited, GAO-15-558R (Washington, D.C.: June 9, 2015).

[20] Executive Order 13514, Federal Leadership in Environmental, Energy, and Economic Performance (Oct. 5, 2009). Executive Order 13653, Preparing the United States for the Impacts of Climate Change (Nov. 1, 2013). Executive Order 13514 was revoked by Executive Order 13693, Planning for Federal Sustainability in the Next Decade, on March 19, 2015.

[21] The Council of Environmental Quality (CEQ) coordinates federal environmental efforts and works closely with agencies and other White House offices in the development of environmental policies and initiatives. CEQ was established within the Executive Office of the President by the National Environmental Policy Act of 1969 (NEPA).

[22] The June 2014 policy statement reaffirmed and superseded the adaptation policy statement the Corps issued on June 3, 2011.

[23] The National Security Council is the President's principal forum for considering national security and foreign policy matters with his senior national security advisors and cabinet officials.

[24] U.S. Army Corps of Engineers, Water Resources Policies and Authorities Incorporating Sea-Level Change Considerations in Civil Works Programs, Engineer Circular 1165-2-211 (Washington, D.C.: July 1, 2009).

[25] U.S. Army Corps of Engineers, Incorporating Sea Level Change in Civil Works Programs, Engineering Regulation 1165-2-212 (Washington, D.C.: Dec. 31, 2011); Incorporating Sea Level Change in Civil Works Programs, Engineering Regulation 1100-2- 8162 (Washington, D.C.: December 31, 2013).

[26] Sea level change can also cause shoreline erosion, inundation or exposure of low-lying coastal areas, shifts in the extent and distribution of wetlands and other coastal habitats, changes to

groundwater levels, and alterations to salinity intrusion into estuaries and groundwater systems. Estuaries are bodies of water formed where freshwater from rivers and streams flows into the ocean, mixing with the sea water.

[27] U.S. Army Corps of Engineers, Procedures to Evaluate Sea Level Change: Impacts, Responses, and Adaptation, Engineering Technical Letter 1100-2-1 (Washington, D.C.: June 30, 2014).

[28] U.S. Army Corps of Engineers, Guidance for Incorporating Climate Change Impacts to Inland Hydrology in Civil Works Studies, Design, and Projects, Engineering and Construction Bulletin 2014-10 (Washington, D.C.: May 2, 2014).

[29] U.S. Army Corps of Engineers, Appropriate Application of Paleoflood Information for Hydrology and Hydraulics Decisions, Engineering Technical Letter 1100-2-2 (Washington, D.C.: Oct. 31, 2014).

[30] A reservoir rule curve is the maximum elevation to which the Corps can fill a reservoir during various times during the year, with the exception of real-time flood operations.

[31] Section 1046 of WRRDA 2014 included a provision for GAO to conduct an audit to determine, among other things, if the Corps' reviews of project operations complied with policies and requirements of applicable law and regulations and to submit a report to Congress by June 10, 2016. We are commencing work on the Corps' project operations, which is to be completed in 2016.

[32] According to the Silver Jackets website, the program provides a formal and consistent strategy for an interagency approach to planning and implementing measures to reduce the risks associated with flooding and other natural hazards. Additional information on the Silver Jackets program is found at http://www.nfrmp.us/state/.

[33] See Department of Defense, Deputy Chief Management Officer, Implementation of Updated Conference Oversight Requirements, Memorandum (Nov. 6, 2013).

[34] David Raff, Levi Brekke, Kevin Werner, Andy Wood, and Kathleen White, Short-Term Water Management Decisions: User Needs for Improved Climate, Weather, and Hydrologic Information (Washington, D.C.: 2012); Levi D. Brekke, Bureau of Reclamation, Technical Service Center, Addressing Climate Change in Long-Term Water Resources Planning and Management: User Needs for Improving Tools and Information (Washington, D.C.: 2011).

[35] GAO, Climate Change: Federal Efforts Under Way to Assess Water Infrastructure Vulnerabilities and Address Adaptation Challenges, GAO-14-23 (Washington, D.C.: Nov. 14, 2013). The Federal Support Toolbox is available at www.watertoolbox.us.

[36] The Corps is required by law to carry out a national program of inspection of dams for the purpose of protecting human life and property. 33 U.S.C. § 467a. The Corps is also required to carry out a levee safety initiative. 33 U.S.C. § 3303a. For these programs, the Corps conducted an inventory of dams and levees before carrying out its risk assessments.

[37] One Corps dam that is newly constructed has not yet undergone an initial screening.

[38] Dam safety modification costs are presented in fiscal year 2013 dollars.

[39] We have work ongoing on the Dam Safety Program that is continuing to assess these issues and will report the results in 2016.

[40] The remaining 80 percent of levee miles are either (1) federally authorized but operated and maintained by local sponsors or (2) in a nonfederal system. Section 3016 of WRRDA 2014 included a provision for GAO to report to Congress on opportunities for alignment of federal programs relating to levees.

[41] Included in the Levee Safety Program are those levees that are: (1) operated and maintained by the Corps, (2) federally authorized but local sponsor operated and maintained, and (3) nonfederal levee systems in the Corps' Rehabilitation and Inspection Program, which

implements Pub. L. No. 84-99's authority to repair and rehabilitate flood control projects damaged by floods and coastal storm events. There are also an unknown number of miles of other levees in the United States that are not included in the Corps' Levee Safety Program. The National Levee Database includes, among other items, the Federal Emergency Management Agency region, the name of the local sponsor(s), the length of the levee(s), a link to a map of the levee(s), and the inspection date and rating. We have work ongoing on the Levee Safety Program that is continuing to assess these issues and will report the results in 2016.

[42] The Corps' Civil Works Transformation Initiative states that that the Corps will be developing a master inventory list of its infrastructure assets. Corps officials said they do not yet know when this inventory will be completed.

[43] Appropriation acts for fiscal year 2006 appropriated funds for and required the Corps to conduct a comprehensive hurricane protection analysis and design of flood control, coastal restoration, and hurricane protection measures for the southeastern Louisiana coastal region. Pub. L. No. 109-103, tit. I, 119 Stat. 2247, 2247-48 (2005), amended by Pub. L. No. 109-148, § 5009, 119 Stat. 2680, 2814 (2005). In addition, one of the appropriation acts appropriated funds for and required the Corps to conduct an analysis and design for comprehensive improvements or modifications to existing improvements in the coastal area of Mississippi in the interest of hurricane and storm damage reduction, prevention of saltwater intrusion, preservation of fish and wildlife, prevention of erosion and other related water resource purposes. Pub. L. No. 109-148, 119 Stat. 2680, 2761 (2005). The Supplemental Appropriations Act for fiscal year 2013 appropriated funds for and required the Corps to conduct a comprehensive study to address the flood risks of vulnerable coastal populations in areas impacted by Hurricane Sandy within the boundaries of the North Atlantic Division of the Corps. Pub. L. No. 113-2, tit. II, 127 Stat. 4, 5 (2013).

[44] U.S. Army Corps of Engineers, Hurricane Sandy Coastal Projects Performance Evaluation Study (Washington, D.C.: November 2013); North Atlantic Coast Comprehensive Study: Resilient Adaptation to Increasing Risk (Washington, D.C.: January 2015).

[45] The Disaster Relief Appropriations Act, 2013, appropriated just over $1 billion for the Corps' Flood Control and Coastal Emergencies Program to prepare for flood, hurricane, and other natural disasters and to support emergency operations, repairs, and other activities after Hurricane Sandy. Pub. L. No. 113-2, div. A, tit. X,127 Stat. 4, 25 (2013).

[46] GAO, Army Corps of Engineers: Organizational Realignment Could Enhance Effectiveness, but Several Challenges Would Have to Be Overcome, GAO-10-819 (Washington, D.C.: Sept. 1, 2010).

[47] A federal court recently ruled that the Army Corps' construction, expansion, operation, and failure to maintain a project in Louisiana caused storm surge during several hurricanes and severe storms that flooded properties, which was a temporary taking under the Fifth Amendment of the United States Constitution. The Takings Clause of the Fifth Amendment prohibits the government from taking private property for a public purpose without just compensation. St. Bernard Parish Gov't v. United States, No. 05-1119, slip op. at 73 (Fed. Cl. 2015). The court has not yet assessed damages, and its decision may be appealed so it is unclear whether, or how, this decision will impact Corps' operations.

[48] Coastal projects, for the purposes of the vulnerability assessment, are those that are within 40 miles of NOAA's tidally influenced water bodies.

[49] U.S. Army Corps of Engineers' business lines include: flood risk reduction, navigation, ecosystem restoration, hydropower, recreation, regulatory, water supply, and emergency management.

In: Water Resources …
Editor: Harry M. Foster

ISBN: 978-1-53610-420-2
© 2016 Nova Science Publishers, Inc.

Chapter 2

LEVEE SAFETY: ARMY CORPS AND FEMA HAVE MADE LITTLE PROGRESS IN CARRYING OUT REQUIRED ACTIVITIES[*]

United States Government Accountability Office

ABBREVIATIONS

Corps	U.S. Army Corps of Engineers
FEMA	Federal Emergency Management Agency

WHY GAO DID THIS STUDY

Levees, which are man-made structures such as earthen embankments or concrete floodwalls, play a vital role in reducing the risk of flooding. Their failure can contribute to loss of lives or property, as shown by the devastation of Hurricane Katrina in 2005. It is estimated that there are over 100,000 miles of levees across the United States, many of which are owned or operated by nonfederal entities. The Corps and FEMA are the two principal federal agencies with authorities related to levee safety.

---

[*] This is an edited, reformatted and augmented version of The United States Government Accountability Office publication, Report to Congressional Requesters GAO-16-709, dated July 2016.

The Water Resources Reform and Development Act of 2014 requires the Corps and FEMA to take the lead on certain national levee-safety-related activities including developing a national levee inventory, which Congress authorized in 2007. The act also includes a provision for GAO to report on related issues. This report examines the Corps' and FEMA's progress in carrying out key national activities related to levee safety required in the act. GAO reviewed pertinent federal laws and executive orders as well as budget, planning, and policy documents from the Corps and FEMA; compared agency activities with federal internal control standards; and interviewed Corps and FEMA headquarters officials.

WHAT GAO RECOMMENDS

GAO recommends that the Corps and FEMA develop a plan that includes milestones for implementing the required national levee-safety-related activities using existing resources or requesting additional resources as needed. The agencies generally concurred with GAO's recommendation.

WHAT GAO FOUND

The U.S. Army Corps of Engineers (Corps) and the Federal Emergency Management Agency (FEMA) have made little progress in implementing key national levee-safety-related activities required in the Water Resources Reform and Development Act of 2014. More specifically, the Corps has been working to develop a national levee inventory, but the agencies have taken no action on the remaining key national levee-safety-related activities for which they are responsible under the act, as shown in the table below. Agency officials identified resource constraints as a primary reason for their lack of progress in implementing such activities, and Corps officials said that not implementing these activities could potentially result in safety risks and federal financial risks for disaster relief, among other impacts. However, the agencies have no plan for implementing the remaining activities required by the act. Without a plan that includes milestones for accomplishing these activities using existing resources or requesting additional resources as needed, the agencies are unlikely to make progress implementing the activities under the act.

Implementation Status, as of June 2016, of Key National Levee-Safety-Related Activities in the Water Resources Reform and Development Act of 2014

Activity	Implementation status	Statutory deadline	Agency responsible
Reconvene the national committee on levee safety	No action	None	Corps
Continue to develop national levee inventory	Ongoing	None	Corps
Implement multifaceted levee safety initiative	No action	Several	Corps and FEMA
Submit a report on the state of U.S. levees, the effectiveness of the levee safety initiative, and any necessary congressional actions	No action	June 10, 2015 and biennially thereafter	Corps
Submit a report including recommendations on advisability and feasibility of a joint dam and levee-safety program	No action	June 10, 2017	Corps and FEMA
Submit a report including recommendations that identify and address legal liabilities of engineering levee projects	No action	June 10, 2015	Corps

Source: GAO analysis of Corps and Federal Emergency Management Agency (FEMA) information. | GAO-16-709.

* * *

July 26, 2016

The Honorable James Inhofe
Chairman

The Honorable Barbara Boxer
Ranking Member
Committee on Environment and Public Works
United States Senate

The Honorable Bill Shuster
Chairman

The Honorable Peter DeFazio
Ranking Member
Committee on Transportation and Infrastructure
House of Representatives

Levees—man-made structures such as earthen embankments or concrete floodwalls—play a vital role in reducing the risk of flooding, and a levee's failure can lead to the loss of lives or property. The impact of levee failures was brought to national attention in 2005, when waves from Hurricane Katrina resulted in the overtopping and failure of levees in parts of New Orleans contributing to major devastation, including at least 1,300 deaths. Addressing the damage from the hurricane cost the federal government more than $16 billion in disaster relief, according to data from the Federal Emergency Management Agency (FEMA). It is estimated that, across the United States, there could be more than 100,000 miles of levees, the majority of which—over 85,000 miles—are owned, maintained, or operated by nonfederal stakeholders, such as states, local governments, tribes, and private entities.[1] FEMA data indicate that levees are found in approximately 22 percent of U.S. counties, where almost half of the U.S. population resides. According to the U.S. Army Corps of Engineers (Corps) and FEMA officials, the federal government does not have a program overseeing all levees across the nation, and no national standards exist for levee safety.

A number of federal agencies have a role in levee-related activities, such as inspections, inventories, assessments, training and assistance, development of standards, mapping, and risk communication; however, the Corps, within the Department of Defense, and FEMA, within the Department of Homeland Security, are the two principal federal agencies with authorities related to levees.[2] The Corps has some responsibility for approximately 15,000 miles of levees under its Levee Safety Program, including levees that the Corps has built and maintains, levees that it has built but does not maintain, and levees that it did not build but rehabilitates.[3] The aim of this program, according to a Corps document, is to better understand, manage, and reduce flood risks associated with levees through such activities as maintaining a national inventory of levees, inspecting and assessing the performance of certain levees, and providing levee rehabilitation to eligible flood-damaged levees. FEMA implements the National Flood Insurance Program, under which public and private levees can be accredited as designed to withstand a certain flood event. The program provided insurance to help protect over 5.1 million policyholders against flood losses in 2015.

In 2007, Congress established the National Committee on Levee Safety to develop recommendations for a national levee safety program.[4] The committee comprised 23 members, including both federal and nonfederal stakeholders. In 2009, the committee recommended, among other things, that federal levee-related programs be aligned to promote levee safety nationwide. More specifically, the committee stated that all federal programs that significantly impact governmental and individual decision making in leveed areas must be aligned toward (1) the goal of reliable levees, (2) an informed and involved public, (3) shared responsibility for the protection of human life, and (4) mitigation of public and private economic damages. In keeping with this goal, the committee called on the Corps, with assistance from FEMA, to develop voluntary national levee-safety guidelines to help ensure best engineering practices for levees are implemented throughout the nation at all levels of government. Further, the committee discussed the potential for federal levee-related programs to be aligned to provide incentives, such as financial or technical assistance, to nonfederal stakeholders to promote more shared responsibility for levee safety. In its report, the committee stated that the average age of levees within federal levee-safety programs was approximately 50 years but that many nonfederal levees could be much older—100 years old or more.

In June 2014, the Water Resources Reform and Development Act of 2014 was enacted[5] and included provisions to, among other things, increase the capacity of nonfederal stakeholders to promote levee safety. The act requires the Corps and FEMA to take the lead on certain national levee-safety-related activities, including establishing voluntary national levee-safety guidelines and providing financial and technical incentives to nonfederal stakeholders to take various actions to promote levee safety. The act also includes a provision for us to submit a report concerning related issues. This report examines the Corps' and FEMA's progress in carrying out key national activities related to levee safety under the Water Resources Reform and Development Act of 2014.

To examine the progress that the Corp and FEMA have made in carrying out key national levee- safety-related activities under the Water Resources Reform and Development Act of 2014, we reviewed pertinent federal laws and executive orders, including the Water Resources Reform and Development Act of 2014; the Water Resources Development Act of 2007;[6] the Moving Ahead for Progress in the 21st Century Act;[7] Executive Order 11988, *Floodplain Management;*[8] *and Executive Order 13690, Establishing a Federal Flood Risk Management Standard and a Process for Further Soliciting and Considering*

Stakeholder Input.[9] We reviewed Corps and FEMA budget, planning, and policy documents concerning the agencies' levee-related programs, as well as documentation about interagency efforts in which they participate. We also reviewed documents developed by the National Committee on Levee Safety, including its 2009 draft report on recommendations for a national levee safety program and its website.[10] In addition, we reviewed our past reports related to levee safety.[11] We also interviewed Corps and FEMA officials who are responsible for implementing the levee-safety-related provisions of the Water Resources Reform and Development Act of 2014. On the basis of our reviews and interviews, we identified the implementation status of key national levee-safety-related activities and compared their status with the requirements in the Water Resources Reform and Development Act of 2014.

We conducted this performance audit from August 2015 to July 2016 in accordance with generally accepted government auditing standards. Those standards require that we plan and perform the audit to obtain sufficient, appropriate evidence to provide a reasonable basis for our findings and conclusions based on our audit objectives. We believe that the evidence obtained provides a reasonable basis for our findings and conclusions based on our audit objectives.

BACKGROUND

This section includes information on the types of levee structures and potential levee failures, major levee-related programs of the Corps and FEMA, and selected legislation related to levee safety.

Types of Levee Structures and Potential Failures

The Water Resources Reform and Development Act of 2014 defines a levee as a manmade barrier (e.g., as an embankment, floodwall, or other structure), the primary purpose of which is to provide hurricane, storm, or flood protection relating to seasonal high water, storm surges, precipitation, or other weather events; such a barrier is normally subject to water loading for only a few days or weeks during a calendar year. According to a Corps document, levees are usually earthen embankments or concrete floodwalls, which have been designed and constructed to contain, control, or divert the flow of water so as to reduce the risk of temporary flooding. An American

Society of Civil Engineers public information document describes earthen levees as being constructed from compacted soil that is typically covered with various surface materials,[12] such as grass, gravel, stone, asphalt, or concrete, to help prevent erosion. The document further states that a floodwall is a vertical levee structure usually erected in urban areas where there is insufficient land for an earthen levee.[13] Levees can either function passively or can require active operations depending on their components. Some levees have gates and pumps, for example, and may require personnel to operate these devices in times of floods. Levees typically require regular maintenance and periodic upgrades to retain their level of protection. Maintenance can include such actions as removing debris and unwanted vegetation from the levees, areas adjacent to floodwalls, and channels; controlling damage caused by animals (e.g., filling burrows); painting or greasing structural components, such as metal gates; and repairing concrete damage, particularly in northern climates with severe freeze-thaw cycles. Figure 1 depicts an earthen levee and a floodwall as well as their respective components.

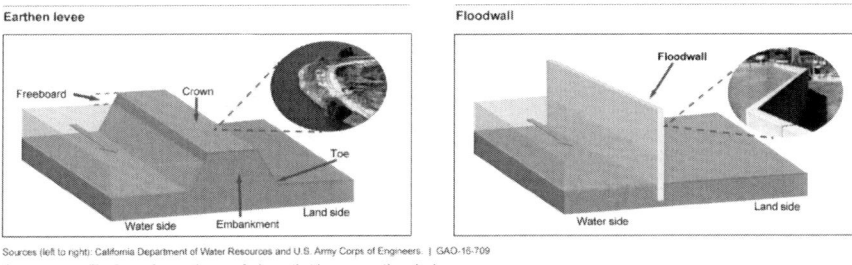

Figure 1. Illustration of an Earthen Levee, a Floodwall, and Their Components.

According to FEMA documents, levees are designed to provide a specific level of protection. However, they can be overtopped or fail; they can also decay over time.

Corps' and FEMA's Major Levee-Related Programs

The Corps and FEMA combined have three primary levee-related programs: the Corps' Levee Safety Program, the Corps' Flood Risk Management program, and FEMA's National Flood Insurance Program.

Corps' Levee Safety Program

According to Corps documents, the Corps' Levee Safety Program, established in 2007, works to better understand, manage, and reduce the flood risks associated with levees through various activities. For example, the Corps maintains a national inventory of levees and makes the information available in the National Levee Database.[14] In addition, the Corps inspects and assesses the performance of about 2,500 levees, comprising about 15,000 miles, nationwide to determine associated risks. On the basis of information from its assessments, the Corps makes recommendations about future federal investments and to prioritize maintenance, repairs, and other actions on levees.

Corps' Flood-Risk Management Program

The Corps' Flood Risk Management Program, established in 2006, is intended to work across multiple Corps' programs to reduce and manage flood risk, according to the Corps' website. The program promotes the appropriate use of levees and floodwalls or alternative actions to reduce flood risk, such as land acquisition and flood proofing. The Corps also communicates levee-related concerns to stakeholders and works with stakeholders to develop solutions to reduce flood risk. The Corps accomplishes this outreach and communication through its flood risk management program as well as through other programs such as the Silver Jackets program, which, according to the Corps' website, is intended to bring together multiple federal, state, and sometimes local agencies and tribes to learn from one another and help reduce the risk of flooding and other natural disasters and enhance response and recovery efforts.

FEMA's NATIONAL FLOOD INSURANCE PROGRAM

FEMA's primary levee-related program is the National Flood Insurance Program, which was first authorized in the National Flood Insurance Act of 1968[15] to, among other things, addresses the increasing cost of federal disaster assistance by providing flood insurance to property owners in flood-prone areas, where such insurance was either not available or prohibitively expensive. This act also authorized subsidies to encourage community and property owner participation. To participate in the program, communities must adopt and agree to enforce floodplain management regulations to reduce the risk of future flood damage. An integral part of the program is the accreditation of any levees near the communities. In exchange for meeting program requirements, federally backed flood insurance is offered to residents in those communities.

Selected Legislation Related to Levee Safety

The Water Resources Development Act of 2007 directed the Corps to create and maintain a National Levee Database that includes a national inventory of levees, with information on the location and condition of all federal levees and, to the extent such information is provided to the Corps, nonfederal levees among other things.[16] It also established the National Committee on Levee Safety to develop recommendations for a national levee safety program. The committee, which was composed of 23 diverse professionals from federal, state, and local or regional governments as well as the private sector and Indian tribes, operated from 2007 to 2011. In 2009, it submitted a draft report to Congress that included 20 recommendations for actions to establish a national levee-safety program, in addition to a strategic plan for implementing the program.[17]

The Moving Ahead for Progress in the 21st Century Act, enacted in 2012, called for the Corps and FEMA to align agency processes to allow interchangeable use of information collected for the Corps' Inspection of Completed Works Program and FEMA's National Flood Insurance Program. In 2013, a joint Corps and FEMA taskforce determined that under certain circumstances, Corps risk assessments of levees conducted under the agency's Levee Safety Program could satisfy aspects of levee accreditation under FEMA's National Flood Insurance Program.[18] The effort culminated in a memorandum of understanding signed by the Corps and FEMA in which the Corps agrees to, among other things, provide FEMA with risk assessment results and FEMA agrees to accept and consider the Corps results, when possible.[19]

The Water Resources Reform and Development Act of 2014 amends portions of the Water Resources and Development Act of 2007 and also requires the Corps and FEMA to take the lead in implementing certain key national levee-safety-related activities. More specifically, it established new reporting responsibilities for the National Committee on Levee Safety, required continued development of a national levee inventory,[20] and required implementation of a multifaceted levee safety initiative under which the agencies are to accomplish the following tasks:

- *Develop voluntary national levee-safety guidelines:* The voluntary national levee-safety guidelines are intended to be comprehensive standards that are available for use by all federal, state, and local agencies as well as tribes. Under the act, the voluntary guidelines are also expected to address activities and practices by states, local governments, tribes, and private entities to safely build, regulate, operate, and maintain a wide range of levee types, canal structures, and related facilities. The guidelines are also expected to address federal activities—including levee inspection, levee rehabilitation, local floodplain management, and public education and training—that facilitate state efforts to develop and implement effective state programs for levee safety.
- *Adopt a hazard potential classification system:* A hazard-potential classification system, as described by the National Committee on Levee Safety in its 2009 draft report, would be a first step in identifying and prioritizing hazards in leveed areas and is to be based solely on the potential consequences associated with a levee's failure, as opposed to the likelihood or probability of a levee failure. The act provides for such a system to be considered in the development of the voluntary national levee-safety guidelines; under the act, the system is also expected to be consistent with the Corps' levee-safety action-classification tool, which ranks levees based on their likelihood of flooding and the associated consequences. According to Corps officials, the tool is currently being used on levees within the Corps' Levee Safety program.
- *Provide technical assistance and materials*: The agencies are to provide technical assistance and training to help promote levee safety and assist states, communities, and levee owners in (1) developing levee safety programs; (2) identifying and reducing flood risks associated with levees; and (3) identifying local actions that may be carried out to reduce flood risks in leveed areas.

- *Provide public education and promote awareness:* To improve public understanding of the role of levees, the agencies are to carry out public education and awareness efforts about the risks associated with living in leveed areas. Education and awareness efforts are to be directed particularly toward individuals living in leveed areas. These efforts must also promote consistency in how information about levee-related risks is communicated at the state and local level and shared among federal agencies.
- *Develop guidelines and provide assistance for a national state and tribal levee-safety program:* This national program, as described by the National Committee on Levee Safety in its 2009 draft report, would assist states and tribes in developing and maintaining the institutional capacity, expertise, and framework to quickly initiate and maintain their own levee-safety program activities and requirements.[21] The guidelines are to identify the minimum components necessary for an individual state or tribe to participate in the program. The national program provides assistance to help establish state and tribal programs that would meet these requirements. The act also requires that state and tribal levee-safety programs will have to adopt the voluntary national levee-safety guidelines to be eligible for assistance.
- *Develop guidelines and provide assistance for a levee rehabilitation assistance program:* This program is to provide assistance to states, local governments, and tribes related to addressing flood mitigation activities that result in an overall reduction of flood risk.[22] The Corps, in consultation with FEMA, is to develop guidelines for floodplain management plans that program participants are required to prepare to reduce the impacts of future floods in areas with levees. Assistance provided under the program may be used for any rehabilitation activity to maximize risk reduction associated with levees that are (1) under a participating state or tribal levee-safety program and (2) not federally operated and maintained. To be eligible, applicants are expected to comply with all applicable federal floodplain management and flood insurance programs, have a floodplain management plan, have a hazard mitigation plan that includes all levee risks, and act in accordance with the voluntary national levee safety guidelines.

In addition, among other things, the act called for several reports to be prepared. Specifically, the Corps is to submit to Congress and make publicly available a biennial report that describes the state of levees in the United States and the effectiveness of the levee safety initiative, as well as any recommendations for legislation and other congressional actions necessary to ensure national levee safety.[23] The Corps and FEMA are also required to submit a report that included recommendations on the advisability and feasibility of, and potential approaches for, establishing a joint national dam and levee safety program, and the Corps is required to submit a report that includes recommendations that identify and address any legal liabilities associated with levee engineering projects.

THE CORPS AND FEMA HAVE MADE LITTLE PROGRESS ON KEY ACTIVITIES UNDER THE ACT, CITING RESOURCE CONSTRAINTS, AND DO NOT HAVE A PLAN FOR IMPLEMENTING THE REST

The Corps and FEMA have made little progress in implementing key national levee-safety-related activities under the Water Resources Reform and Development Act of 2014 primarily because of resource constraints, according to officials from both agencies. The Corps has been working on its development of a national levee inventory, but the Corps and FEMA have not begun work on other key national leveesafety-related activities required by the act and do not have a current plan for doing so (see table 1).

Table 1. Implementation Status, as of June 2016, of Key National Levee-Safety-Related Activities in the Water Resources Reform and Development Act of 2014

Activity	Implementation status	Statutory deadline	Agency responsible
Reconvene the national committee on levee safety	No action	None	Corps
Continue to develop national levee inventory	Ongoing	None	Corps
Implement multifaceted levee safety initiative			
Develop voluntary national levee-safety guidelines	No action	June 10, 2015	Corps and FEMA
Establish a hazard potential classification system	No action	None	Corps

Activity	Implementation status	Statutory deadline	Agency responsible
Provide technical assistance and training	No action	None	Corps and FEMA
Provide public education and promote awareness	No action	None	Corps and FEMA
Issue guidelines that establish minimum components for state and tribal levee-safety program	No action	June 10, 2015	Corps and FEMA
Provide assistance for a state and tribal levee-safety program	No action	None	FEMA
Develop guidelines for preparation of floodplain management plans under the levee assistance programs	No action	Dec. 7, 2014	Corps and FEMA
Provide assistance for a levee rehabilitation assistance program	No action	None	Corps
Submit report on the state of U.S. levees, the effectiveness of the levee safety initiative, and any necessary congressional actions	No action	June 10, 2015, and biennially thereafter	Corps
Submit report, including recommendations, on advisability and feasibility of a joint dam and levee safety program	No action	June 10, 2017	Corps and FEMA
Submit report including recommendations that identify and address legal liabilities of engineering levee projects	No action	June 10, 2015	Corps

Source: GAO analysis of U.S. Army Corps of Engineers (Corps) and Federal Emergency Management Agency (FEMA) information. | GAO-16-709.

Concerning the national levee inventory, a summary document that the Corps developed for us states that the Corps is incorporating levee data that FEMA has provided from the National Flood Insurance Program and is working to incorporate levee data voluntarily provided by state and local agencies. The Corps' actions are an extension of earlier work on the database, which it was directed to establish and maintain under the Water Resources and

Development Act of 2007. Corps officials said that improving the inventory will be an ongoing process. The Corps had allocated $5 million for the inventory in fiscal year 2016, and the Corps' fiscal year 2017 Operations and Maintenance budget justification lists an allocation of an additional $5 million to further expand the inventory.

The agencies have taken no action on the remaining key national levee-safety-related activities for which they were responsible and have missed several statutory deadlines for developing guidelines and reports. For example, the agencies took no action on developing the guidelines for the preparation of floodplain management plans under the levee rehabilitation assistance program, which were due on December 7, 2014; the voluntary national levee-safety guidelines, due June 10, 2015; or a report, due June 10, 2015, that was to include, among other things, recommendations for legislation and other congressional actions necessary to ensure national levee safety. Additionally, according to agency officials we interviewed, the agencies have no current plan for implementing the remaining activities. Without a plan, including milestones for accomplishing these activities using existing resources or requesting additional resources as needed, the agencies are unlikely to make further progress on implementing the remaining activities required by the act.

Corps officials we interviewed said that they have continued to make progress on other activities that will complement activities required by the Water Resources Reform and Development Act of 2014 and that are within the scope of their existing Levee Safety Program and Flood Risk Management Program. Similarly, FEMA officials stated that they also are working to provide general public education and promote awareness about the risks associated with living behind levees through their existing National Flood Insurance Program.

In a slide presentation that the Corps prepared for us, dated October 2015, the Corps identified resource constraints as a primary reason why the Corps has not been able to carry out certain key national leveesafety-related activities under the Water Resources Reform and Development Act of 2014. Specifically, the Corps' presentation indicated that new appropriations would be needed to (1) provide technical assistance and training; (2) develop guidelines and provide financial assistance for a state and tribal levee-safety program; and (3) develop guidelines and provide financial assistance for a levee rehabilitation assistance program.[24] Corps officials we interviewed stated that the remaining national levee-safety-related activities required in the act could be funded using existing appropriations, but these activities would have to compete with existing Corps projects in the Corps civil works program.[25] We reviewed a 2016 Corps budget document and determined that, except for the national inventory of levees, the Corps did not specifically allocate funds for national levee-safety-related activities required in the act.

FEMA officials we interviewed stated that the agency would need additional appropriations to carry out the agency's main responsibility under the act—providing assistance for a state and tribal levee safety program—and told us that the agency had not received any funding directed toward national activities required by the act. They also said that even if these activities were funded, the agency would need additional staffing resources—specifically, in its 10 regional offices—to carry out requirements under the act. As of this report, FEMA has one staff person who is available part-time to implement the national levee-safety-related activities required by the act.

As noted above, the Corps' 2017 budget includes $5 million for the national levee inventory; however, it does not specify funds for implementing the other national levee-safety-related activities in the Water Resources Reform and Development Act of 2014. Corps headquarters officials told us that not implementing the act's national levee-safety-related activities could result in several potential impacts, including that the disaster relief burden for the federal government may increase, safety risks and loss of life may increase, and risk education in communities with levees may not be carried out.

Conclusion

Since the devastation of Hurricane Katrina in 2005, Congress has enacted legislation, including the Water Resources Reform and Development Act of 2014 that provided the Corps and FEMA with lead responsibility for undertaking certain national levee-safety-related activities, including some that would increase the capacity of nonfederal stakeholders to promote levee safety. The Corps is working on one of the key national levee-safety-related activities required by the act, namely expanding a national inventory of levees. However, the Corps and FEMA have not taken action to implement the other activities, required by the act, citing resource constraints. Further, Corps officials have identified potential impacts—including safety and financial risks—of not carrying out these activities, but the agencies do not have a plan for implementing these activities. Without a plan, including milestones for accomplishing the activities using existing resources or requesting additional resources as needed, the agencies are unlikely to make further progress implementing the activities under the act.

RECOMMENDATION FOR EXECUTIVE ACTION

To help ensure that the Corps and FEMA carry out the national levee-safety-related activities required in the Water Resources Reform and Development Act of 2014, we recommend that the Secretary of Defense direct the Secretary of the Army to direct the Chief of Engineers and Commanding General of the U.S. Army Corps of Engineers and that the Secretary of Homeland Security direct the FEMA Administrator to develop a plan, with milestones, for implementing these activities, using existing resources or requesting additional resources as needed. This plan could be posted on the Corps' website and monitored for progress.

AGENCY COMMENTS

We provided a draft of this report for review and comment to the Departments of Defense and Homeland Security. In their written comments, respectively, both agencies generally concurred with our recommendation. The Department of Defense stated that the agencies are drafting an implementation plan and suggested that we focus our recommendation on finalization of this plan. However, the agencies did not provide a copy of the draft plan or a date when it would be finalized, so we believe that the current focus of the recommendation is appropriate. The Department of Defense further stated that, to date, no funding has been allocated to the Corps specifically to implement provisions under the Water Resources Reform and Development Act of 2014, except for the levee inventory activities, as we have acknowledged in our report. In addition, the Department of Defense suggested that the recommendation be revised to include posting the plan on the Corps' website and monitoring the plan for progress. We have modified our recommendation to incorporate this suggestion, which we believe would help inform nonfederal stakeholders who own, maintain, or operate the majority of levees. The Department of Homeland Security said that FEMA will continue to work with the Corps to develop and implement a plan to carry out key national safety-related activities required in the act. Both agencies also provided technical comments that we incorporated, as appropriate.

Anne-Marie Fennell
Director,
Natural Resources and Environment

End Notes

[1] This estimate, which is the most recent available, is based on information contained in a 2009 report from the National Committee on Levee Safety. The committee was established by Congress in 2007 and is no longer active.

[2] Other federal agencies with roles related to levee safety include the Natural Resources Conservation Service in the Department of Agriculture; the National Oceanic and Atmospheric Administration in the Department of Commerce; the Department of Housing and Urban Development; the Bureau of Reclamation, U.S. Fish and Wildlife Service, and U.S. Geological Survey in the Department of the Interior; and the Environmental Protection Agency.

[3] The Corps, through its Rehabilitation and Inspection Program, supplements local efforts to repair flood risk management projects, including levees, after they are damaged during a flood. Pub. Law No. 84-99, 69 Stat. 1866 (1955). To be eligible for rehabilitation assistance under this program, the project must meet specified design and construction criteria and the levee sponsor must maintain the levee to specified standards.

[4] Pub. Law No. 100-114 §9003, 121 Stat. 1041, 1288.

[5] Pub. Law No. 113-121, 128 Stat. 1193.

[6] Pub. Law No. 100-114.

[7] Pub. Law No. 112-141, 126 Stat. 405 (2012).

[8] Exec. Order No. 11,988, 42 Fed. Reg. 26,951 (May 24, 1977).

[9] Exec. Order No. 13,690 (Washington D.C.: Jan. 30, 2015).

[10] Draft: Recommendations for a National Levee Safety Program: A Report to Congress from the National Committee on Levee Safety (Jan. 15, 2009). The report was sent to Congress in draft form.

[11] GAO, Army Corps of Engineers: Efforts to Assess the Impact of Extreme Weather Events, GAO-15-660 (Washington, D.C.: July 22, 2015) and FEMA and the Corps Have Taken Steps to Establish a Task Force, but FEMA Has Not Assessed the Costs of Collecting and Reporting All Levee-Related Concerns, GAO-11-689R (Washington, D.C.: July 29, 2011).

[12] American Society of Civil Engineers, So, You Live Behind a Levee! (Reston, Virginia: 2010).

[13] Floodwalls are sometimes constructed on the crown of an earthen levee to increase the levee's height and its ability to control larger floods.

[14] According to a Corps document, the National Levee Database incorporates the best available data on the location and condition of levees and flood walls nationwide and then displays that information in an interactive map on its website. The database helps facilitate the linkage among levee safety activities, such as flood risk communication, levee evaluations for the National Flood Insurance Program, levee inspections, floodplain management, and risk assessments.

[15] Pub. L. No. 90-448, Tit. XIII, 82 Stat. 476, 572 (codified as amended at 42 U.S.C. §§ 4001-4129).

[16] Pub. Law No.100-114 §9004 (a)(2).

[17] Draft Report to Congress from the National Committee on Levee Safety.

[18] United States Army Corps of Engineers and Federal Emergency Management Agency, Flood Protection Structure Accreditation Task Force: Final Report (November 2013).

[19] Memorandum of Understanding between the Federal Emergency Management Agency and the United States Army Corps of Engineers for Alignment of Levee Activities, Information, and Messaging (Nov. 13, 2014).

[20] The Corps was initially charged with establishing and maintaining a national levee inventory as part of the National Levee Database under the Water Resources Development Act of 2007.

[21] Draft Report to Congress from the National Committee on Levee Safety.

[22] This program is separate from Corps' existing Rehabilitation and Inspection Program, authorized in 1955, Pub. Law No. 84-99, 69 Stat. 1866, to supplement local efforts to repair flood-risk management projects, including levees, after they are damaged during a flood.

[23] Under the act, this report is to be done in coordination with the National Committee on Levee Safety and is to be done biennially after the first report is issued.

[24] The latter two programs account for $55 million of the $79 million authorized by the act. However, no funds were specifically appropriated for these programs in fiscal years 2015 to 2016. Unlike many other federal agencies that have budgets established for broad program activities, most of the Corps' civil works funds are directed for specific projects. The conference report accompanying the appropriation acts generally lists individual projects and specific allocations of funding for each project.

[25] According to Corps documents, the Corps' Civil Works Program develops, manages, restores, and protects the nation's water resources through studies of potential projects, construction of projects, operation and maintenance of projects, and its regulatory program. Through the program, the Corps also works with other federal agencies to help communities respond to and recover from floods and other natural disasters.

In: Water Resources …
Editor: Harry M. Foster

ISBN: 978-1-53610-420-2
© 2016 Nova Science Publishers, Inc.

Chapter 3

ARMY CORPS OF ENGINEERS: ACTIONS NEEDED TO IMPROVE COST SHARING FOR DAM SAFETY REPAIRS[*]

United States Government Accountability Office

ABBREVIATIONS

ASA (CW)	Assistant Secretary of the Army for Civil Works
Corps	U.S. Army Corps of Engineers
DSAC	Dam Safety Action Classification
DOD	Department of Defense
DSO	Dam Safety Officer
IWTF	Inland Waterways Trust Fund
PMA	Power Marketing Administration
Reclamation	Bureau of Reclamation
SEPA	Southeastern Power Administration
WRDA	Water Resources Development Act of 1986

---

[*] This is an edited, reformatted and augmented version of The United States Government Accountability Office publication, Report to Congressional Requesters GAO-16-106, dated December 2015.

Why GAO Did This Study

The Corps operates over 700 dams, which are aging and may require major repairs to assure safe operation. At some dams, sponsors that benefit from dam operations share in the cost of operating and repairing these dams based on original congressional authorizations for dam construction or subsequent agreements with the Corps. Since 2005, the Corps initiated an estimated $5.8 billion in repairs at 16 dams with urgent repair needs; sponsors are to share repair costs at 9 of these dams.

GAO was asked to examine cost sharing for Corps dam safety repairs. This report examines how, over the last 10 years, the Corps (1) determined cost sharing and (2) communicated with sponsors regarding cost sharing. GAO reviewed relevant laws and Corps regulations; analyzed dam safety projects' documentation for the 16 dams the Corps selected for repairs since 2005; conducted site visits to a non-generalizable sample of three dams based on cost share determinations and range of sponsors; and interviewed Corps officials and sponsors.

What GAO Recommends

GAO recommends that the Corps clarify policy guidance on (1) usage of the state-of-the-art provision and (2) effective communication with sponsors to establish and implement cost sharing agreements for all dams, including the three named in this report. The Department of Defense concurred with GAO's recommendations.

What GAO Found

The U.S. Army Corps of Engineers (Corps) determined sponsors' (such as water utilities and hydropower users) share of costs for dam safety repairs pursuant to regulations, but did not apply a provision in a statutory authority that reduces sponsors' share. The Corps determined these cost shares based on analyses of the potential ways each dam could fail, and in consideration of statutory requirements regarding which type of cost sharing arrangement, or authority, would apply given these possible failure scenarios.

- The Corps applied its *Major Rehabilitation* authority at 11 of the 16 dam safety repair projects GAO reviewed for repairs associated with typical degradation of dams, such as embankment or foundation erosion through seepage. Under this authority, sponsors are to pay their full agreed-upon cost share of the repair.
- The Corps applied its *Dam Safety Assurance* authority at 5 of the 16 dam safety repair projects GAO reviewed for repairs that resulted from the availability of new hydrologic or seismic data. Under this authority, sponsors' agreed-upon cost share is reduced by 85 percent.

The Corps did not apply one provision of its *Dam Safety Assurance* authority— related to repairs needed due to changes in state-of-the-art design or construction criteria (state-of-the-art provision)—since the enactment of the enabling legislation in 1986. Since that time, the Corps has not provided guidance on the types of circumstances under which the state-of-the-art provision applies and has not had a consistent policy position regarding the provision. For example, the Corps' latest regulation states in one section that the state-of-the-art provision will not be applied because of the difficulty in defining terminology, while another section allows for consideration on a case-by-case basis. Without clarifying the circumstances under which the state-of-the-art provision applies, and implementing the policy consistently, the Corps is at risk of not applying the full range of statutory authorities provided to it, contributing to conditions under which, as discussed below, sponsors have taken actions opposing the Corps.

In GAO's review of 9 dams with sponsors, the Corps did not communicate with or effectively engage all sponsors. For example, a federal sponsor that markets hydropower generated at two dams disagreed with the Corps' decision to not apply the state-of-the-art provision of its *Dam Safety Assurance* authority, which, if used, would reduce this sponsor's cost share by about $410 million. This sponsor has proceeded to set its power rates in anticipation of paying the reduced cost share, creating uncertainty for the recovery of federal outlays for repairs. In addition, GAO found the Corps was not effective in reaching agreement with other sponsors on cost-sharing responsibilities at three dams because it did not have clear guidance for effectively communicating with sponsors. For example, the Corps did not engage a sponsor to ensure cost share payment at one dam and, at another dam, delayed executing agreements that would ensure sponsors' cost shares. Because the Corps did not effectively engage these sponsors, some are deriving benefits absent agreements with the Corps, while others that have agreements have not been notified of their final cost-sharing responsibility. As a result, these sponsors' cost share payments (about $3.1 million) are uncertain.

* * *

December 10, 2015

The Honorable Barbara Boxer
Ranking Member
Committee on Environment & Public Works
United States Senate

The Honorable Bill Shuster
Chairman

The Honorable Peter DeFazio
Ranking Member
Committee on Transportation and Infrastructure
House of Representatives

The Honorable David Vitter
United States Senate

The U.S. Army Corps of Engineers (Corps) operates 709 dams[1] that provide numerous benefits for a wide range of customers, including protecting communities from floods, generating hydropower, and supplying water from reservoirs. While the Corps' dams comprise a small portion of the country's more than 87,000 dams,[2] they are a part of the aging national infrastructure.[3] For example, the American Society of Civil Engineers estimates that by 2020, 70 percent of all dams in the United States will be over 50 years old.[4] Currently, the average age of the Corps' 709 dams is 56 years.[5]

The age and criticality of dam infrastructure requires the Corps to conduct regular maintenance and, in some cases, major repairs to assure continued safe operation. The Corps currently estimates the cost of fixing all of its dams that need repair at $24 billion. Since 2005, when the Corps adopted its current risk-informed approach to dam safety, it has initiated repairs of 16 dams in urgent need of repair, the costs for which range from tens to hundreds of millions of dollars per dam, with total repair costs estimated at about $5.8 billion. At some dams, sponsors or organizations, such as water utilities or hydropower users, that benefit from dam operations share in the cost of the repairs. In this context, you asked us to review issues concerning cost sharing for dam safety repairs. This report examines how, over the last 10 years, the Corps (1) determined cost sharing for dam safety repairs and (2) communicated with sponsors regarding cost sharing for dam safety repairs.

To address these objectives, we reviewed relevant federal laws and Corps engineering regulations related to dam safety and cost sharing. For each of the 16 dam safety repair projects funded for design or construction from fiscal year 2007[6] to fiscal year 2016, we analyzed Corps dam safety documents and compared them against the Corps' latest *Safety of Dams* regulation.[7,8] We interviewed Department of Defense and Corps headquarters officials about how the Corps determined and communicated with sponsors about cost sharing for dam safety repairs. In particular, we interviewed officials about the Corps' process for dam safety repairs, key decision points for determining cost sharing, communication with sponsors, and tracking of cost share payments. We interviewed Corps officials at the 11 district offices where the 16 dams are located about their decisions associated with cost sharing and about their communication with sponsors. We also interviewed the federal, state, local, and private sponsors identified by the Corps about their cost sharing in these dam safety repair projects (see app. I for the list of sponsors we interviewed). We asked the sponsors about the terms of their agreements with the Corps, their history of being a sponsor, the financial impacts of cost sharing for dam safety repair projects, and the Corps' communication with them regarding the projects and cost sharing. We compared communications between sponsors and the Corps against requirements for such communications described in the Corps' latest *Safety of Dams* regulation. Additionally, we conducted site visits to a non-generalizable sample of 3 dams in the Corps' Nashville, TN, and Tulsa, OK, districts based on the Corps' cost sharing determinations and the range of project sponsors (e.g., hydropower, water supply). At these sites we observed dam safety repair projects and interviewed Corps officials and sponsors.

We conducted this performance audit from November 2014 to December 2015 in accordance with generally accepted government auditing standards. Those standards require that we plan and perform the audit to obtain sufficient, appropriate evidence to provide a reasonable basis for our findings and conclusions based on our audit objectives. We believe that the evidence obtained provides a reasonable basis for our findings and conclusions based on our audit objectives.

BACKGROUND

The Corps is the world's largest public engineering, design, and construction management agency, responsible for water resources infrastructure such as dams, levees, hurricane barriers, and floodgates in every state.[9] Through its Civil Works program, the Corps plans, designs, and operates water resources infrastructure projects. The Civil Works program is organized into 3 tiers: a national headquarters in Washington, D.C.; 8 regional divisions that were established generally according to watershed boundaries; and 38 districts nationwide. In addition, the Corps maintains national and regional centers that provide technical services to Corps divisions and districts, such as support of dam safety repair projects.

The Assistant Secretary of the Army for Civil Works (ASA (CW)), appointed by the President, establishes the strategic direction, develops policy, and supervises the execution of the Civil Works program. The Corps headquarters and regional division offices primarily implement policies and provide oversight to district offices. The Corps headquarters' Dam Safety Officer (DSO), a civilian official, is responsible for all dam safety activities, including establishing policy and technical criteria for dam safety and prioritizing dam-safety-related work. The eight divisions, commanded by military officers, coordinate civil works projects in the districts within the eight respective geographic areas. The Corps districts, commanded by military officers, are responsible for planning, engineering, constructing, and managing water resources infrastructure projects in their districts as well as coordinating with the Corps' sponsors.

Most of the Corps' dams are one of two types: earthen or concrete.[10] According to Corps data, about 68 percent of Corps dams have earthen embankments, constructed of various types of materials such as clay, silt, sand, or gravel. Another 30 percent of Corps dams are concrete dams.[11] Dams can have various features, such as spillway gates and conduit outlets, to control water releases, as well as auxiliary spillways to divert water flows in the event of expected maximum flood conditions. (See fig. 1.)

Army Corps of Engineers

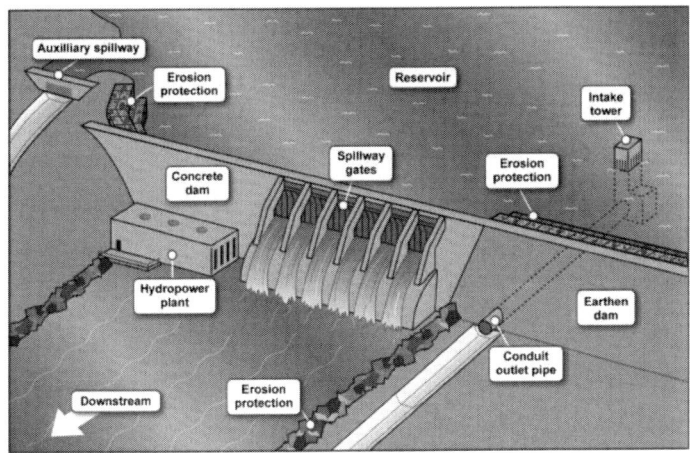

Source: GAO. | AO-16-106.

Figure 1. Illustration of Dam Types and Features.

The Corps' Dam Safety Repair Process

To ensure continued safe operation, Corps dams undergo routine maintenance, such as cleaning of drains and mowing of embankments, but in some cases require major repairs, which, as defined by the Corps, are those that cost over $16 million. These repairs may be to:

- rehabilitate spillway gate equipment to safely pass excess water,
- fill voids in embankments or foundations with grout,[12]
- build cutoff walls to prevent erosion to embankments or foundations from seepage,[13]
- build shear walls to increase dam stability,[14]
- increase dam's height to prevent overtopping, or
- anchor a dam to its foundation.[15]

Since 2005, the Corps has used a risk-informed approach to select dams for safety-related repairs. While integrating traditional engineering analyses and standards, the risk-informed approach aims to identify and prioritize the most critical dam safety risks rather than eliminate all potential risks. To that end, the Corps has developed the Dam Safety Action Classification (DSAC) system, based on a 5-point scale, to help guide key decisions for dam safety repairs. This risk classification system reflects the probability of a dam's failure and resulting potential consequences due to failure.[16] As of July 2015, the Corps has placed 309 dams (about 44 percent) in actionable categories (DSAC 1, 2, and 3) because the dams were determined to be at moderate to very high risk of failure. In particular, the Corps has classified 17 dams as DSAC 1 (very high urgency), 76 dams as DSAC 2 (high urgency), and 216 dams as DSAC 3 (moderate urgency).[17] From fiscal year 2007 to fiscal year 2016, the Corps selected 16 of these DSAC 1 and 2 dams for repairs.

According to the Corps' *Safety of Dams* regulation,[18] once a dam has been selected as needing repair according to its DSAC designation, the Corps is to take the following steps to study, design, and construct a dam safety repair project.[19]

- *Study:* Corps district officials are to conduct a dam safety modification study to determine a long-term solution. This study is to involve risk analyses, determination of potential failure modes, evaluation of alternatives to address potential failures, and development of a recommended technical solution with its estimated cost.[20] The study also is to identify cost share sponsors and to recommend an applicable authority for cost sharing purposes (discussed later in this report) under which to implement the repair work. The results of the study are published in a dam safety modification report, which is forwarded to division and headquarters officials, including the DSO, for review and approval of recommended repairs. The Corps districts are to communicate to sponsors and the public about dam failure risks and potential repairs during the study phase. Once approved by Corps' DSO and ASA(CW), the cost estimate in the dam safety modification report is used as a basis to request funds from Congress for design and construction.
- *Design:* Project design takes place at the Corps districts and dam safety production centers,[21] involving investigation of site conditions, such as testing soils, engineering analysis, and development of design plans and specifications. In addition, further risk analyses are to be conducted as well as expert reviews of the design. During the project's design, the Corps districts are also to communicate to sponsors and the public about their plans for conducting repairs.
- *Construction:* Project construction, managed by district officials, is typically carried out through contracts with private companies. Construction for dam safety repairs can take multiple years and

involve several contracts. To assure construction quality, the Corps districts are required to conduct regular inspections. In addition, Corps officials are to continue their outreach and communications with sponsors and the public throughout the construction period.

Cost Sharing Sponsors

Sponsors share in the costs of dam safety repairs based on original congressional authorizations for dam construction or subsequent sponsors' agreements with the Corps. A wide array of entities can be cost sharing sponsors, including federal, state, and local agencies as well as private entities. Sponsors may be identified at the time of original dam construction or at a later time. Congressional authorizations or sponsors' agreements with the Corps delineate the benefits sponsors receive as well as their responsibilities and cost sharing obligations. Cost sharing terms are unique to each sponsor at each dam. Commensurate with benefits derived from use of a dam, sponsors typically pay a percentage of a dam's annual operations and maintenance costs, as well as the same percentage of total costs of major dam safety repairs. Cost sharing percentages can range from under 1 percent, such as for small water supply users, to over 50 percent, such as for hydropower users, depending on a sponsor's agreement with the Corps.

Sponsors' payment mechanisms for dam safety repairs vary. When the Corps determines a need for dam safety repairs, it typically budgets for and funds the entire amount of the repair upfront. Sponsors, responsible for sharing in the design and construction costs for dam safety repair projects, pay their cost shares in different ways as described below and in table 1. However, not all Corps dams have cost sharing sponsors. The federal government fully funds the repairs of those Corps dams that do not have sponsors.[22]

- Non-federal sponsors, depending on their agreement with the Corps, are to pay their cost share either on a "pay-as-you-go" basis or at the end of the project. Sponsors that are identified at the time of initial dam construction typically pay their cost share on a pay-as-you-go basis. In these situations, sponsors contribute their cost share while project design and construction are ongoing. Sponsors—typically water utilities—that enter into agreements with the Corps subsequent to the dam's initial construction have the option to pay as you go or in lump sum, with interest, at the end of the dam safety repair project, once all costs are finalized and calculated. According to Corps officials, non-federal sponsors may seek an exception to amortize their cost share payments over time following project completion. The Corps collects and tracks payments submitted by non-federal sponsors and transmits them to the U.S. Treasury.

- Federal sponsors of Corps dams are the U.S. Department of Energy's four Power Marketing Administrations (PMA).[23] PMAs sell the electrical output of federally owned and operated hydroelectric dams.[24] PMAs market wholesale power by entering into contracts with customers, with preference given to not-for-profit public-owned utilities, to sell power at set rates.[25] Through their rates, PMAs recover all costs associated with power production and transmission, including their cost share for dam safety repairs, which they remit directly to the U.S. Treasury. PMAs are to recover all associated power production costs within a reasonable period of time, which the Department of Energy has traditionally considered to be 50 years or less.

Table 1. Cost Sharing Payment Mechanisms and Recipients for Federal and Non-Federal Sponsors of the U.S. Army Corps of Engineers' Dam Safety Repair Projects

	Non-federal sponsors		Federal sponsors
	Original agreement (at time of dam construction)	Subsequent agreement (after dam construction)[a]	Power Marketing Administrations
Payment mechanism	Pay as you go	Pay as you go Lump sum upon project completion	At project completion; payment can be over a period of up to 50 years
Payment recipient	U.S. Treasury through Corps	U.S. Treasury through Corps	U.S. Treasury directly

Source: GAO analysis of U.S. Army Corps of Engineers' documentation. | GAO-16-106.

[a] In general, section 1203 of the Water Resources Development Act of 1986 provides that payment of costs may be made over a period of up to 30 years from the date of a project's completion, with interest, for projects covered by section 1203. According to Corps officials, under Corps policy, while such an amortized payment is available to sponsors that do not have an existing agreement with the Corps, sponsors that do have an existing agreement with the Corps are required to pay using pay as you go or lump sum upon project completion options. Corps officials further provided that they allow for sponsors to seek an exception to this policy.

THE CORPS HAS DETERMINED COST SHARING BASED ON WAYS IN WHICH A DAM MAY FAIL, BUT HAS NOT APPLIED ONE PROVISION THAT REDUCES SPONSORS' COST SHARE

A Dam's Potential Failure Mode Drives the Corps' Decision on Cost Sharing

According to the Corps' *Safety of Dams* regulation.[26] during a dam safety modification study, Corps district officials are to identify and analyze all the potential ways that a dam could fail. Such potential failure modes can include: (1) embankment or foundation erosion through seepage; (2) inability of a dam to safely pass excess water during expected maximum flood conditions (hydrologic failure mode); or (3) inability of a dam to withstand the expected maximum earthquake (seismic failure mode). Once potential failure modes, among other things, are determined, Corps district officials are to generate a dam safety modification report that reviews alternatives and recommends a technical solution to address the potential failure modes.

For cost sharing purposes, the regulation requires the district to recommend in the report one of the two types of cost sharing arrangements or authorities: Major Rehabilitation authority or Dam Safety Assurance authority. The potential failure mode is the primary factor in determining the applicable authority, in addition to consideration of policy and statutory requirements:

Major Rehabilitation: According to Corps officials, this authority applies to dam safety repairs associated with typical degradation of dams over time. Under this authority, sponsors are to pay their full cost share. For example, if a sponsor's agreed cost share is 10 percent, then the sponsor is responsible for 10 percent of the total cost of the dam safety repair project. (See table 2.) The Corps' regulation requires application of Major Rehabilitation authority if embankment or foundation erosion through seepage or instability is determined to be the potential failure mode.

Dam Safety Assurance: In certain situations, however, the Corps can apply its Dam Safety Assurance authority, which significantly reduces sponsors' cost shares. This authority, based on Section 1203 of the Water Resources Development Act (WRDA) of 1986, applies to safety-related dam modifications needed as a result of new hydrologic or seismic data or changes in state-of-the-art design or construction criteria deemed necessary for safety purposes (state-of-the-art provision).[27] This authority reflects, in part, the availability of new information—such as current hydrologic models or seismic studies—that could indicate a dam's increased vulnerability and greater risk of failure. Application of this authority reduces a sponsor's responsibility to 15 percent of its agreed cost share, effectively reducing a sponsor's cost share obligation by 85 percent. For example, if a sponsor's agreed cost share is 10 percent, then the sponsor is responsible for 15 percent of this amount, meaning that it would be responsible for 1.5 percent of the total cost of a dam safety repair project. (See table 2).

Table 2. Hypothetical Cost Sharing Example of a $50-Million Dam Safety Repair Project with 10 Percent Sponsor Cost Share

Sponsor's cost share	Major Rehabilitation	Dam Safety Assurance
Cost sharing responsibility based on agreement	Full (100 percent)	Reduced (15 percent)
Percent of total project cost	10	1.5
Amount of cost share	$5 million	$750,000

Source: GAO analysis | GAO-16-106.

The final determination of cost sharing authority is reviewed through the Corps' chain of command. The Corps' DSO is to review and approve the dam safety modification report and determination of funding authority. Subsequently, the ASA (CW) office is to review the DSO decision and determine if it concurs.[28] Sponsors have no formal role in the Corps' authority determination. According to Corps officials, while the sponsors are typically involved in cost sharing discussions, funding authority determination is a federal responsibility and not subject to appeals from sponsors.

The Corps Consistently Determined Cost Sharing Based on Potential Failure Mode

The Corps applied either its Major Rehabilitation or Dam Safety Assurance authority to the 16 dams selected for dam safety repairs from fiscal year 2007 to fiscal year 2016, selecting the funding authority to address each dam's determined potential failure mode consistent with its regulation. (See app. II.) The total estimated cost for these repairs is $5.8 billion.

- For 11 of the 16 dams the Corps applied its Major Rehabilitation authority. At 9 of these 11 dams, the potential failure mode was determined to be embankment or foundation erosion through seepage, and the Corps implemented dam safety repair projects under its Major Rehabilitation authority consistent with its regulation.[29] Sponsors for these dams are to pay their full cost share, estimated at $574 million of the total $4.2 billion in repairs.[30]

- For the 5 remaining dams, the Corps applied its Dam Safety Assurance authority because repairs were determined to be the result of new hydrologic or seismic data indicating the potential inability of these dams to safely pass excess water during expected maximum flood conditions or to withstand the expected maximum earthquake. The sponsors for these dams are to pay 15 percent of their cost share—which cumulatively total an estimated $31 million of the total $1.6 billion in repairs for these dams.[31]

The Corps Did Not Apply One Provision of Its Dam Safety Assurance Authority That Reduces Sponsors' Cost Share

While the Corps applied the Dam Safety Assurance authority to 5 of 16 dams in our review based on the availability of new hydrologic or seismic data, it did not apply the Dam Safety Assurance authority's state-of-the-art provision to any of these dam safety repair projects. According to ASA(CW) officials, the Corps has not applied the state-of-the-art provision since enactment of the enabling legislation (WRDA of 1986).

When asked why the Corps had not applied this provision, ASA (CW) officials said that they would consider applying the state-of-the-art provision on a case-by-case basis, but they have never been presented with a case that they determined to have merited it. Additionally, ASA (CW) officials were unable to define the conditions under which the provision could apply or to provide a hypothetical example of a dam safety issue that would lead them to use it.

The circumstances under which the state-of-the-art provision might apply have not been identified in the Corps regulations, and the Corps has not had a consistent policy position regarding when the state-of-the-art provision might apply. The Corps' 1997 regulation states that dam safety repairs required due to state-of-the-art changes would be decided on a case-by-case basis, but does not identify criteria for how the cases would be selected. However, in 2011, and again in the 2014 update, the Corps' *Safety of Dams* regulation discusses application of Dam Safety Assurance authority only with regard to new hydrologic or seismic data, stating that the state-of-the-art provision would not be applied. Specifically, the 2014 regulation notes the difficulty of defining the state-of-the-art provision and states that because the state-of-the-art "terminology makes it difficult to define the kinds of repairs that would be applicable, [...] it is not used."[32] The same 2014 regulation states that use of the state-of-the-art provision must be decided on a case-by-case basis by the ASA (CW).

Internal control standards state that information and effective communication are needed for an agency to achieve all of its objectives.[33] Moreover, internal controls guidance states that effective communication may be achieved through clear policy. However, the Corps' current regulation is not clear as to what is meant by "state-of-the-art design or construction criteria deemed necessary for safety purposes" in the statutory provision. Thus, this lack of clarity coupled with the Corps' inconsistent policy position has hindered the Corps from applying the state-of-the-art provision in a manner consistent with other Dam Safety Assurance provisions. Without clarifying the circumstances under which the state-of-the-art provision applies and implementing the policy consistently, the Corps is at risk of not applying the full range of statutory authorities provided to it, thereby raising questions about the appropriate allocation of federal and non-federal funding for dam safety repairs. As discussed later in this report, the Corps' inaction in setting a clear policy for a provision under which sponsors face significant financial impacts has contributed to conditions under which sponsors have asserted their own terms for use of the provision or are considering taking legal action against the Corps.

In contrast, another federal agency has applied a similar state-of-the-art provision to its dam safety repairs.[34] The U.S. Department of the Interior's Bureau of Reclamation (Reclamation) has a similar statutory authority enacted by the Reclamation Safety of Dams Act of 1978, which requires sponsors' cost share at 15 percent when modifications result from new hydrologic or seismic data, or changes in state-of-the-art design or construction criteria deemed necessary for safety purposes.[35] According to Reclamation officials, while Reclamation has not developed a definition for the state-of-the-art design or construction criteria, it has operationalized and applied the state-of-the-art provision exclusively to modify 30 dams since 1978, primarily in situations where defensive dam safety measures, such as filters and drainage mechanisms, were lacking or were not consistent with the current state of the practice.[36]

SOME CORPS DISTRICTS DID NOT COMMUNICATE WITH SPONSORS OR ENGAGE THEM EFFECTIVELY, POTENTIALLY REDUCING PAYMENTS RECEIVED FROM SPONSORS

The Corps Did Not Effectively Communicate Its Cost Sharing Determination, Contributing to Uncertainty Regarding Sponsor Payment

The Corps' lack of clarity and a consistent policy position regarding the state-of-the-art provision under the Dam Safety Assurance authority has contributed to disagreements with a major sponsor and uncertainty regarding sponsor payment. In this case, Southeastern Power Administration (SEPA)[37] the federal PMA sponsor for Center Hill (Tennessee) and Wolf Creek (Kentucky) dams, has disagreed with the Corps' decision to repair the dams under its Major Rehabilitation authority rather than the state-of-the-art provision of the Dam Safety Assurance authority. (See fig. 2.) SEPA has asserted that the Dam Safety Assurance authority should apply to these projects.[38] SEPA has taken this position, in part, because while dam safety repairs at Wolf Creek were originally determined to be under the Major Rehabilitation authority,[39] Corps district officials had subsequently recommended using the Dam Safety Assurance authority based on application of the state-of-the-art provision.[40] SEPA was aware of the district's recommendation to change the authority determination to Dam Safety Assurance. However, the ASA (CW) ultimately did not support this recommendation noting that erosion caused by seepage—the potential failure mode identified at these dams—has consistently and categorically been addressed through application of the Major Rehabilitation authority. According to SEPA officials, the conflicting actions of Corps district and headquarters officials on authority determination created uncertainty for SEPA regarding the Corps' position.

SEPA stated that the need for repairs to Center Hill and Wolf Creek dams is based on state-of-the-art design and construction practices and notes that the Corps consulted with recognized international experts to design the cutoff walls being built at these dams to address the effects of seepage. According to SEPA officials, current repairs based on state-of-the-art practices are being made at these two dams, in part, because previous repair efforts did not adequately address site conditions contributing to seepage.[41] Conversely,

Corps officials told us that seepage naturally occurs at all dams and periodically needs to be addressed, such as through implementation of repair projects. Moreover, according to Corps officials, the "karst" limestone upon which the Center Hill and Wolf Creek dams are built is prone to increasing seepage over time because of the dissolution of soluble rock foundation. Concrete cutoff walls put in place at Center Hill and Wolf Creek dams under current projects were designed to consider these effects and, according to Corps officials, constructed as permanent seepage control measures.

Source: U.S. Army Corps of engineers. | GAO-16-106.

Figure 2. Center Hill, TN, (left) and Wolf Creek, KY, Dams.

Because of the high cost of repairs to these two dams—estimated at about $958 million, for which SEPA's share under its original congressional authorization is about 50 percent—SEPA officials have expressed concern about the agency's ability to recover costs if the projects are considered under the Major Rehabilitation authority. Under this authority, SEPA's cost to recover for both dams is estimated at about $482 million. Officials said that if SEPA were obligated to recover this amount, its hydropower rates could become prohibitively expensive. As a result, according to these officials, SEPA's customers might terminate their contracts and acquire energy via more economical options, such as energy derived from natural gas or coal. If the Corps were to apply its Dam Safety Assurance authority to these repairs under, for example, the state-of-the-art provision, SEPA's cost to recover would be reduced to about $72 million (85 percent reduction).

The outcome related to the disagreement between the Corps and SEPA has significant implications given that mitigating the effects of seepage, as evidenced by our review, is a common reason for making safety-related repairs. In recent rate-making notices, SEPA has based its proposed rates on the Dam Safety Assurance authority for dam safety repairs at Center Hill and

Wolf Creek dams. This action signals SEPA's position that it should pay the reduced cost share (about $72 million) provided under this authority, and without resolution, recovering federal outlays for funding the majority of project costs (about $410 million) remains uncertain. In moving forward to resolve this disagreement, it is important that potential impacts on aging dam infrastructure, hydropower rates, and the federal budget are considered in a coordinated, strategic approach.

SEPA's rate actions could set precedent and create uncertainty for the federal government if sponsors at other dams also assert that the state-of-the-art provision applies to projects that mitigate the effects of seepage. For example, the Corps determined that repairs to mitigate the effects of seepage were needed at 9 of the 16 dams we reviewed, with a total estimated cost of about $4 billion. If other sponsors at these dams were to follow SEPA's example, the federal government could potentially receive reduced cost share payments from these sponsors. Further, in light of its aging infrastructure, more Corps dams could require seepage-related repairs in the future. A policy that clarifies the Corps' application of the state-of-the-art provision could help to minimize potential disagreements with sponsors and lead to greater certainty concerning the federal government's and project sponsors' cost sharing obligations.

The Corps Has Not Engaged Some Sponsors Effectively, Potentially Reducing Sponsors' Payment to the Federal Government

The Corps' *Safety of Dams* regulation requires Corps districts to engage sponsors by notifying them during the study phase about the dam safety repair project and their estimated financial responsibility. The regulation further states: "Requirements for cost sharing and the identification of non-Federal sponsors (or partners) must occur very early in the study process to ensure that the non-Federal interests are willing cost share partners. Uncertainty about sponsorship and the lack of meaningful sponsor involvement in the scope and extent of dam safety repairs can cause delays to the dam safety modification work." As mentioned previously, under the Corps' regulations, Corps district officials are also expected to communicate with sponsors throughout project design and construction as well as officially notify sponsors of their final cost share payment upon the project's completion. Additionally, internal control standards state that managers should effectively communicate with external stakeholders that may have a significant impact on the agency achieving its goals.[42]

While the Corps *Safety of Dams* regulation identifies when communication with sponsors is to occur, it does not provide clear guidance on how to effectively communicate with sponsors to establish and implement cost sharing agreements. Based on our discussions with state, local, and private sponsors of the dams we reviewed, we found that the Corps has generally established good relationships with these nonfederal sponsors and communicated project status information; however, some Corps districts were not timely or effective in communicating and reaching agreement on cost sharing responsibilities. Of the 16 dam safety repair projects we reviewed, 9 had sponsors,[43] and—as discussed below—at 3 of the 9 dams the Corps did not communicate with the sponsors in a manner that would ensure their meaningful involvement and willingness to be cost sharing partners, as required by its regulation. According to the agreements, these sponsors are to pay their cost share to the Corps, which remits these funds to the U.S. Treasury. However, at least three sponsors have expressed concerns and indicated resistance about paying their determined cost shares, estimated to be about $3.1 million. Because the Corps does not have clear guidance to ensure effective communications with sponsors, it did not adequately communicate or reach agreements on cost sharing responsibilities with these sponsors. As a result, these sponsors' plans for paying their cost shares are uncertain, leaving the recovery of federal outlays from these sponsors similarly uncertain.

- *Tuttle Creek Dam:* At Tuttle Creek dam (Kansas), the Corps identified and contacted one water supply sponsor during the study phase (2000–2002) of a dam stabilization project as well as notified the sponsor of its estimated cost share, but otherwise did not effectively engage the sponsor throughout the project to ensure the sponsor's cost share payment. In a 2002 letter to the Corps, the sponsor asserted its position that it should not be required to pay for repairs to stabilize the dam, a repair that would enable the dam to withstand the expected maximum earthquake. In the sponsor's opinion, the sponsor was not responsible for sharing costs related to changes in the Corps' design standards or to address what the sponsor felt were design flaws. In 2003, the Corps responded to the sponsor reiterating the sponsor's responsibility for sharing in the costs of the project. The Corps' written response included its estimate of the sponsor's cost share, approximately $770,000, and described payment options: pay-as-you-go or lump sum at the end of construction. According to the sponsor, it did not raise any further objections and, in a subsequent telephone conversation with Corps district officials, indicated its preference to use the pay-as-you-go option because it would be unable to afford a lump sum payment. Since 2003, the sponsor received briefings on the status of the project; however, the Corps did not follow up or otherwise engage the sponsor to pay incrementally while construction was ongoing. Construction was completed in 2010, but as of October 2015, the Corps had not requested payment or notified the sponsor of its final cost share. Corps officials told us that they are preparing a billing letter to send to the sponsor.

- *Rough River Dam:* At Rough River dam (Kentucky), the Corps' 2012 dam safety modification report stated that the project to grout and construct a 1,700-foot cutoff wall would be completed at full federal expense with no cost sharing sponsors. However, subsequent reviews by Corps headquarters identified water supply contract holders, and in 2013, the Corps notified three water supply sponsors of their cost sharing responsibilities for the dam safety repair. Due to uncertainty in identifying sponsors and delays in executing agreements with them, as discussed below, the Corps may experience challenges collecting these sponsors' cost shares when the project is finally complete, estimated to be no later than 2021. Specifically:
- One sponsor has had a water use agreement with the Corps since 1978, but has not been drawing water from the reservoir since 2007. In 2013, the Corps requested that the sponsor remove its water intake structure from the reservoir. However, in the same year, as mentioned previously, the Corps notified this sponsor of its cost sharing responsibilities for the dam safety repair project. In May 2015, the Corps signed a termination agreement with the sponsor under which the sponsor will not share the costs of the project. While the Corps is not expecting to collect a cost share, its interaction with the sponsor indicates a lack of effective communication.
- Although the Corps notified a second sponsor of its cost sharing responsibilities in 2013, this sponsor currently does not have a cost sharing responsibility for the dam safety repair project because the sponsor paid upfront for "major capital replacement" as part of its 1966 agreement with the Corps.[44] This provision of the agreement is to expire in April 2016, and according to Corps officials, a supplement to the agreement is being developed. The supplement

would include this sponsor's cost sharing responsibility in the current project. However, we were not able to reach this sponsor to confirm its intention to be a cost sharing sponsor, and it remains uncertain whether the Corps should expect a future agreement to cover current project costs.
- The third sponsor has been drawing water from the reservoir since 2002, when the sponsor negotiated terms of its water use with the Corps under a draft contract. Despite drawing up to 1.6 million gallons per day from the reservoir, the sponsor has not paid the Corps for water use and operations and maintenance expenses because a contract between the parties has not been executed. As a result, despite notifying the sponsor of its cost sharing responsibilities in the dam safety repair project in 2013, the Corps has no mechanism to compel payment from this sponsor. According to the sponsor, it has tried to finalize the 2002 contract numerous times, but the Corps did not finalize the agreement in any of these instances. In July 2015, a Corps district official told us that Corps headquarters is reviewing the negotiated agreement; however, uncertainty about cost sharing exists until all parties execute a contract.
- *Center Hill Dam:* At Center Hill Dam (Tennessee), the Corps identified three water supply sponsors during the study phase but generally had minimal interactions with them to communicate cost sharing estimates and responsibilities. While two sponsors accept their cost sharing responsibilities and estimated cost sharing amounts, one sponsor disagrees with the Corps' application of the Major Rehabilitation authority. Similar to the argument made by SEPA, which is also a sponsor at this dam, this water supply sponsor stated that the repairs being made to address the effects of seepage at the dam incorporate state-of-the-art design and construction practices and that the Corps should apply the state-of-the-art provision, thereby reducing this sponsor's cost share. Under the Major Rehabilitation authority, this sponsor has a $1.9 million cost share. According to this sponsor, a municipal water utility, covering this cost would require raising water rates approximately 50 cents per household per month. The sponsor is contemplating a legal challenge if the Corps does not apply the state-of-the-art provision to lower this sponsor's cost share according to a sponsor official. The Corps has maintained its position that application of its Major Rehabilitation authority is appropriate for this dam safety repair.

CONCLUSION

Considering the significant cost of dam safety repair projects, and the number of dams that could need repairs in the future, implementing a dam safety program as effectively as possible is important. This implementation would include adequately defining conditions for key policy determinations to ensure the appropriate allocation of federal versus non-federal funds for dam safety repairs.

However, the fact that the Corps has not developed policy guidance on the types of circumstances under which the state-of-the-art provision of its Dam Safety Assurance authority might apply, and has not had a consistent policy position, limits the Corps' ability to ensure the effective implementation of the dam safety program. Without clarifying the circumstances under which the state-of-the-art provision applies and implementing the policy consistently, the Corps is at risk of not applying the full range of statutory authorities provided to it. Moreover, because of the financial implications of its authority determinations for sponsors, the Corps' inaction in setting a clear policy for this provision contributes to conditions under which it is potentially exposed to adverse actions of these sponsors.

The Corps' engagement of project sponsors is critical to the successful implementation of dam safety repair projects not only to ensure the continued provision of benefits, such as water supply and hydropower generation, but also to recover federal outlays used to fund projects upfront. Because the Corps has not always effectively communicated with or engaged sponsors, some are deriving benefits from dams absent an agreement with the Corps while other sponsors that have agreements either have not been notified by the Corps of their final cost share responsibility or dispute the Corps' cost sharing determination and may raise a legal challenge. While the Corps' *Safety of Dams* regulation provides guidance to district offices for communicating with sponsors, greater clarity about effective communication requirements to establish and implement agreements with sponsors would help the Corps ensure equity in its treatment of sponsors and make certain that the federal government receives expected cost share payments.

RECOMMENDATIONS

To improve cost sharing for dam safety repairs, we recommend that the Secretary of Defense direct the Secretary of the Army to direct the Chief of Engineers and Commanding General of the U.S. Army Corps of Engineers to clarify policy guidance:

- on the types of circumstances under which the state-of-the-art provision of the Dam Safety Assurance authority might apply to dam safety repair projects.
- for district offices to effectively communicate with sponsors to establish and implement cost sharing agreements during dam safety repair projects. For all dams, including the three dams named in the report, this would involve communicating estimated and final cost sharing amounts, executing agreements, and engaging sponsors to ensure cost share payment.

AGENCY COMMENTS

We provided a draft of this report to the Department of Defense (DOD) for official review and comment. In its written comments, DOD concurred with our recommendations and described the actions it plans to take within the next 18 months.

In response to our recommendation to clarify policy guidance on the types of circumstances under which the state-of-the-art provision of the Dam Safety Assurance authority might apply, the department stated that the ASA(CW) will clarify the usage of the provision within the next 18 months.

Regarding our recommendation to clarify policy guidance for district offices to communicate with sponsors to establish and implement cost sharing agreements, DOD stated that ASA(CW) will review and clarify policy, guidance, and business practices related to communication with sponsors within the next 18 months. With respect to the three dam safety repair projects identified in our report, the department stated that the ASA(CW) will engage with their sponsors to establish a path forward to recouping the federal investment in the Corps' work, including finalization of water supply agreements.

Lori Rectanus
Director, Physical Infrastructure Issues

APPENDIX I:
LIST OF SPONSORS INTERVIEWED

The table below lists all sponsors we interviewed for this report. Not all sponsors for the dams included in our review were available for interview. Additionally, because the Southeastern Power Administration (SEPA) is a major cost sharing sponsor, we interviewed the Tennessee Valley Public Power Association, an organization that represents 155 local utilities across seven states that purchase wholesale power marketed by SEPA.

Table 3. List of 10 Sponsors Interviewed

Dam, State	Sponsor(s)
Bolivar and Dover, OH	Muskingum Watershed Conservancy District
Canton Lake, OK	Oklahoma City Utilities Department
Center Hill, TN	City of Cookeville Department of Water Control, DeKalb Utility District, City of Smithville, Southeastern Power Administration
Isabella, CA	Kern River Watermaster[a]
Pine Creek, OK	International Paper Company
Rough River, KY	Grayson County Water District
Tuttle Creek, KS	State of Kansas (Kansas Water Office)
Wolf Creek, KY	Southeastern Power Administration

Source: GAO | GAO-16-106.

[a] The Kern River Watermaster represents five interests: (1) Kern Delta Water District, (2) Buena Vista Water District, (3) City of Bakersfield, (4) North Kern Water Storage District, and (5) Kern County Water Agency.

Appendix II: List of 16 Dam Safety Repair Projects Reviewed

These projects received funding for design and construction from fiscal year 2007 to fiscal year 2016.

#	Dam name, state	Potential failure mode	Repair	Project sponsor	Cost sharing agreement's %	Authority	Final/latest project cost	Estimated cost sharing amount
1	Addicks & Barker Dams, TX	Embankment erosion through seepage	• Construct new outlet structures with three gated conduits to pass water at Addicks and Barker • Grout and abandon in place existing outlet structures at Addicks and Barker • Construct 1,400-foot long cutoff wall along Barker	None	N/A	MR	$129,883,000	N/A
2	Bluestone Dam, WV	Inability to safely pass excess water during expected maximum flood conditions	• Stabilize dam with 564 anchors and concrete blocks • Construct 330-foot wide auxiliary spillway	None	N/A	DSA	$527,300,000	N/A
3	Bolivar Dam, OH	Foundation erosion through seepage	• Construct 4,500-foot underground concrete wall along embankment • Replace 6 service gates and rehabilitate associated machinery	Muskingum Watershed Conservancy District	23.00%	MR	$109,284,000	$25,135,320
4	Canton Lake Dam, OK	• Inability to safely pass excess water during expected maximum flood conditions • Inability to withstand expected maximum earthquake	• Stabilize dam with 64 anchors • Construct 480-foot wide auxiliary spillway	Oklahoma City Utilities Department	25.50%	DSA	$183,800,000	$7,030,350*
5	Center Hill Dam, TN	Embankment erosion through seepage	• Grout and construct 1,000-foot concrete cutoff wall along main dam embankment • Construct 900-foot wide reinforcing berm downstream of auxiliary dam embankment	Southeastern Power Administration	42.545%	MR	$364,200,000	$154,948,890
				City of Cookeville Department of Water Control	0.5330%	MR		$1,941,186
				City of Smithville	0.0320%	MR		$116,544
				Dekalb Utility District	0.0530%	MR		$193,026
				North Alabama Bank	0.0100%	MR		$36,420

#	Dam name, state	Potential failure mode	Repair	Project sponsor	Cost sharing agreement's %	Authority	Final/latest project cost	Estimated cost sharing amount
6	Clearwater Lake Dam, MO	Embankment erosion through seepage	Grout and construct 4,200-foot concrete cutoff wall along embankment	None	N/A	MR	$211,440,000	N/A
7	Dover Dam, OH	Inability to safely pass excess water during expected maximum flood conditions	• Raise dam by 8 feet along 860-foot length • Stabilize dam with 140 anchors	Muskingum Watershed Conservancy District	23.00%	DSA	$60,000,000	$2,070,000[a]
8	East Branch Dam, PA	Embankment erosion through seepage	Grout and construct 2,100-foot concrete cutoff wall within the embankment	None	N/A	MR	$248,000,000	N/A
9	Emsworth Dam, PA	• Gate failure • Failure of erosion protection	• Install 14 new gates and gate hoisting systems • Install 1,700 feet of erosion protection across both dams and 120 feet downstream of both dams	Inland Waterways Trust Fund	50%	MR	$160,000,000	$48,146,000[b]
10	Herbert Hoover Dike, FL	Embankment erosion through seepage	• Install internal erosion protection through entire length of embankments (80 miles) • Replace 28 and remove 4 water control structures	None	N/A	MR	$2,069,510,000	N/A
11	Isabella Dam, CA	Main and auxiliary dams: • Inability to safely pass excess water during expected maximum flood conditions Auxiliary dam: • Inability to withstand expected maximum earthquake and fault rupture • Embankment erosion through seepage	Main dam: • Raise dam by 16 feet along 2,000-foot embankment Auxiliary dam: • Raise dam by 16 feet along 3,500-foot embankment • Construct 80-foot wide downstream buttress along 3,500-foot embankment • Construct new emergency spillway 900-foot wide	North Kern Water Storage District; Buena Vista Water Storage District; Kern Delta Water District; City of Bakersfield; Kern County Water Agency	21%	DSA	$680,771,000	$21,444,287[a]
12	Lockport Dam, IL	Embankment erosion through seepage	• Construct 4,300-foot slurry trench wall along right embankment • Construct 1,200-foot roller-compacted concrete wall on right embankment to replace existing wall • Rehabilitate 2.5 miles of existing concrete cutoff wall along left embankment	Inland Waterways Trust Fund	50%	MR	$149,175,575	$14,400,000[b]
13	Pine Creek Dam, OK	Embankment erosion into and along conduit pipe	• Construct 124-foot concrete cutoff wall • Install 480-foot steel pipe liner for conduit	International Paper Company	8%	MR	$29,900,000	$2,392,000
14	Rough River Dam, KY	Embankment erosion through seepage	Grout and construct 1,700-foot concrete cutoff wall along embankment	Grayson County	0.0868%	MR	$147,000,000	$127,596
				City of Leitchfield	0.0380%	MR		$55,860
15	Tuttle Creek Dam, KS	Seepage and piping after embankment slope failure caused by a seismic event	• Construct 351 shear walls • Rehabilitate equipment for 18 gates	State of Kansas, Kansas Water Office	2.49%	DSA	$166,700,000	$622,625[a]
16	Wolf Creek Dam, KY	Embankment erosion through seepage	Grout and construct 4,000-foot concrete cutoff wall along embankment	Southeastern Power Administration	55.113%	MR	$593,710,821	$327,211,844

Legend: MR = Major Rehabilitation authority; DSA = Dam Safety Assurance authority
Source: GAO analysis based on U.S. Army Corps of Engineers data. | GAO-16-106.

[a] The amount reflects an 85 percent reduction in cost share due to application of the Dam Safety Assurance authority.

[b] The cost share for this project was reduced due to congressional actions in fiscal year 2009.

c A portion of this project's cost was shared with the Inland Waterways Trust Fund (IWTF). Prior to fiscal year 2014, the Corps contributed 100 percent to the project. Fiscal year 2014 allocations are 50 percent Corps and 50 percent IWTF.

End Notes

[1] A dam is an artificial barrier constructed for the purpose of storage, control, or diversion of water. The Corps defines dams as being (1) 25 feet or more in height or (2) having an impounding capacity at maximum water storage elevation of 50 acre-feet or more.

[2] The Corps maintains and publishes the National Inventory of Dams, which contains information about dams in the United States and its territories.

[3] According to the American Society of Civil Engineers, in 2013, the majority of dams, 69 percent, were owned by a private entity. Federal, state, and local governments owned and operated the remaining dams. According to Corps' data, Corps dams represent about 50 percent of all federally-owned dams.

[4] American Society of Civil Engineers. 2013 Report Card for America's Infrastructure (March 2013).

[5] According to Corps data, most Corps dams were built during the 1960s and 1970s, with 75 percent of dams serving multiple purposes related to flood control, irrigation, navigation, water supply, hydropower generation, and recreation. More than half of the Corps' dams are located on or east of the Mississippi River.

[6] Fiscal year 2007 was the first fiscal year that reflected the Corps' current risk-informed dam safety approach initiated in 2005. The Corps' risk-informed approach is described in more detail later in the report.

[7] We did not evaluate the accuracy or legal sufficiency of the Corps' application of cost-sharing formulas and calculations.

[8] U.S. Army Corps of Engineers, Safety of Dams—Policy and Procedures, Regulation No. 1110-2-1156 (Mar. 31, 2014).

[9] Located within the Department of Defense, the Corps has both military and civilian responsibilities. The Corps' Military program provides, among other things, engineering and construction services to other U.S. government agencies and foreign governments. This report discusses the Civil Works program.

[10] Some of the Corps' dams are combination earthen and concrete dams. The Corps categorizes dams by their primary type.

[11] The remaining 2 percent of the Corps' dams are other types.

[12] Grout is a fluidized material injected into soil, rock, concrete, or other construction material to seal openings and to lower the permeability and/or provide additional structural strength.

[13] A cutoff wall is a wall of impervious material, usually of concrete, asphaltic concrete, or steel sheet piling constructed in the foundation and abutments to reduce seepage beneath and adjacent to the dam.

[14] A shear wall is a structural element used to resist lateral forces parallel to the plane of the wall. Normally, a series of walls are built at set intervals along the downstream foundation of a dam to resist movement, or separation, of a dam from its foundation.

[15] Multi-strand cables connecting the dam to its foundation can be installed and placed in tension to anchor the dam and prevent its displacement.

[16] Dam failure is characterized by the sudden, rapid, and uncontrolled release of impounded water. Possible consequences of dam failure include loss of life and property.
[17] The Corps has categorized 398 dams as DSAC 4 (low urgency) and none as DSAC 5 (normal urgency). As of July 2015, two Corps dams were not classified: one newly constructed and one newly added to the inventory.
[18] U.S. Army Corps of Engineers, Safety of Dams—Policy and Procedures, Regulation No. 1110-2-1156 (Mar. 31, 2014).
[19] Some dam safety repair projects in our review did not fully follow the risk-informed process because they were initiated prior to the process being instituted.
[20] A recommended solution may not require a repair. To reduce the risk of dam failure, the Corps may, for example, lower the reservoir impounded by the dam.
[21] Regional Dam Safety Production Centers assign a lead engineer to support technical development of a project. In addition, the Dam Safety Modification Mandatory Center of Expertise—a national center of expertise—provides technical support to the project.
[22] According to the Corps, non-federal interests that are not project sponsors may provide contributions such as granting rights-of-way or easements in support of repair projects.
[23] The four PMAs are: The Bonneville Power Administration (BPA), the Western Area Power Administration (WAPA), the Southeastern Power Administration (SEPA), and the Southwestern Power Administration (SWPA).
[24] Under federal statute, power generated at Corps dams beyond what is needed for dam operations is to be delivered to the Secretary of Energy who is to transmit and dispose of such power in a manner as to encourage the most widespread use at the lowest possible rates to consumers consistent with sound business principles (16 U.S.C. § 825s).
[25] To set rates, PMA Administrators propose draft rates, which are approved on an interim basis by the Deputy Secretary of Energy, and notify the public of proposed 5-year rates through Federal Register Notices. Rate schedules become effective upon confirmation and approval by the Secretary of Energy.
[26] U.S. Army Corps of Engineers, Safety of Dams—Policy and Procedures, Regulation No. 1110-2-1156 (Mar. 31, 2014).
[27] Pub. L. No. 99-662, §1203, 100 Stat. 4082, 4263 (1986) (codified at 33 U.S.C. § 467n).
[28] While authority determination is a Corps responsibility, on two occasions Congress has specifically authorized or directed the Corps to apply its Dam Safety Assurance authority. These involved projects at Fern Ridge dam, OR (Pub. L. No. 110-114, § 5120, 121 Stat. 1041, 1240 (2007)) and Beaver Lake dam, AR (Pub. L. No. 102-377, 106 Stat. 1315, 1318 (1992), Pub. L. No. 102-580, § 209(f), 106 Stat. 4797, 4830 (1992)).
[29] The potential failure modes of the remaining two dams were gate failure and erosion along a conduit pipe.
[30] Seven of the 11 dams have sponsors. The remaining 4 dams are 100 percent federally funded.
[31] Four of the 5 dams have sponsors. The remaining dam is 100 percent federally funded.
[32] U.S. Army Corps of Engineers, Safety of Dams—Policy and Procedures, Regulation No. 1110-2-1156 (Mar. 31, 2014).
[33] GAO, Standards for Internal Control in the Federal Government, AIMD-00-21.3.1 (Washington, D.C.: November 1999).
[34] We did not evaluate the effectiveness of this agency's dam safety repair efforts or the extent to which it consistently applied the applicable state-of-the-art provision as part of this review.
[35] 43 U.S.C. § 508(b) enacted by Pub. L. No. 95-578, § 4(b), 92 Stat. 2471 (1978).

[36] As of July 2015, Reclamation solely applied the state-of-the-art provision to safety repairs at 30 dams. It applied the state-of-the-art provision in combination with a hydrologic and/or seismic provision at an additional 21 dams.

[37] SEPA markets hydropower generated by Corps dams to not-for-profit public-owned utilities in the states of Georgia, Florida, Alabama, Mississippi, Illinois, Virginia, Tennessee, Kentucky, North Carolina, and South Carolina.

[38] See SEPA's notice of proposed rates at 80 Fed. Reg. 30451 (May 28, 2015). In addition, on October 2, 2015, SEPA issued a notice of interim approval of rates based upon the application of section 1203 of the Water Resources Development Act of 1986 (80 Fed. Reg. 59742 (Oct. 2, 2015)). In the notice of interim approval of rates, SEPA notes, however, that as it continues to finalize its rate calculation, it also continues to discuss, analyze, and seek guidance from other relevant agencies. SEPA further provides that interagency discussions on funding authority remain ongoing and that a reconsideration of such interim rates could occur if, as a result of those discussions, other relevant federal agencies provide a factual and legal basis for a contrary determination regarding the applicability of section 1203 of the Water Resources Development Act of 1986.

[39] See U.S. Army Corps of Engineers, Wolf Creek Dam, Jamestown, Kentucky: Seepage Control Major Rehabilitation Evaluation Final Report (July 11, 2005).

[40] Wolf Creek dam was the first of the two dams to undergo dam safety repairs. Both Center Hill and Wolf Creek dams are located in the Cumberland River basin and, according to Corps documents, have the same underlying geology and dam safety concerns. SEPA has argued that Dam Safety Assurance authority similarly applies to repairs at both dams.

[41] SEPA officials cited the Corps' July 2005 report (see footnote 39) as the basis for their comments. According to this report, original design and construction techniques of the 1930s and 1940s used at Wolf Creek were inadequate to control seepage in the "karst" geology beneath the dam. In addition, according to this report, installation of a grout curtain and cutoff wall from 1968 to 1979 failed to adequately limit seepage.

[42] GAO/AIMD-00-21.3.1.

[43] While 11 of the 16 dams in our review had sponsors, only 9 dams had sponsors that were organizations. The Inland Waterways Trust Fund, the sponsor for 2 remaining dams, is a funding source financed through a fuel tax for construction and rehabilitation of locks and dams on the Inland Waterways System.

[44] In its agreement with the Corps, the sponsor paid $56 upfront for "major capital replacement" required during the 50-year term of the agreement (i.e., until 2016).

In: Water Resources …
Editor: Harry M. Foster

ISBN: 978-1-53610-420-2
© 2016 Nova Science Publishers, Inc.

Chapter 4

ARMY CORPS OF ENGINEERS: ADDITIONAL STEPS NEEDED FOR REVIEW AND REVISION OF WATER CONTROL MANUALS*

United States Government Accountability Office

ABBREVIATIONS

Corps	U.S. Army Corps of Engineers
NOAA	National Oceanic and Atmospheric Administration
O&M	operations and maintenance
USGS	U.S. Geological Survey

WHY GAO DID THIS STUDY

The Corps owns and operates water resource projects, including more than 700 dams and their associated reservoirs across the country, for such purposes as flood control, hydropower, and water supply. To manage and operate each project, the Corps' districts use water control manuals to guide project operations. These manuals include water control plans that describe the

---

* This is an edited, reformatted and augmented version of The United States Government Accountability Office publication, Report to Congressional Requesters GAO-16-685, dated July 2016.

policies and procedures for deciding how much water to release from reservoirs. However, many of the Corps' projects were built more than 50 years ago, and stakeholders have raised concerns that these manuals have not been revised to account for changing conditions.

The Water Resources Reform and Development Act of 2014 included a provision for GAO to study the Corps' reviews of project operations, including whether practices could better prepare the agency for extreme weather. This report (1) examines the extent to which the Corps has reviewed or revised selected water control manuals and (2) describes the Corps' efforts to improve its ability to respond to extreme weather. GAO reviewed the Corps' guidance on project operations; examined agency practices; and interviewed Corps officials from headquarters, all 8 divisions, and 15 districts—selected, in part, on regional differences in weather conditions.

WHAT GAO RECOMMENDS

GAO recommends that the Corps develop guidance on what constitutes a water control manual's review and how to document it and track which manuals need revision. The agency concurred with the recommendations.

WHAT GAO FOUND

According to U.S. Army Corps of Engineers (Corps) officials, the agency conducts ongoing, informal reviews of selected water control manuals and has revised some of them, but the extent of the reviews and revisions is unclear because they are not documented or tracked, respectively. The Corps' engineer regulations state that water control manuals should be reviewed no less than every 10 years so that they can be revised as necessary. However, officials from all 15 districts GAO interviewed said they do not document informal reviews of water control manuals because they consider such reviews part of the daily routine of operating projects. The Corps does not have guidance, consistent with federal standards for internal control, on what activities constitute a review or how to document the results of reviews. Without such guidance, the Corps does not have reasonable assurance that it will consistently conduct reviews and document them to provide a means to retain organizational knowledge. The Corps' engineer regulations also state that water control manuals shall be revised as needed, but the extent to which manuals have been revised or need revision remains unknown because the Corps' divisions do not track consistent information about manuals. For example, based on GAO's review of the Corps' documents, one of the eight divisions tracked whether the water control plans in its water control manuals reflected actual operations of a project, but the remaining seven did not. While the Corps has revised certain water control manuals as called for by its regulations, district officials GAO interviewed said additional manuals need revision. However, the Corps does not track consistent information on manuals needing revision, in accordance with federal internal control standards. Without tracking which manuals need revision, it is difficult for the Corps to know the universe of projects that may not be operating in a way that reflects current conditions as called for in the Corps' engineer regulations.

The Corps has efforts under way to improve its ability to respond to extreme weather, including developing a strategy to revise drought contingency plans and studying the use of forecasting to make decisions on project operations. To better respond to drought, the Corps is developing a strategy to analyze drought contingency plans in its water control manuals to account for a changing climate. As of May 2016, the Corps was conducting, as a pilot, updates of five projects' drought contingency plans to help test methods and tools for future use in other plans. The Corps is also studying the use of forecasting tools to improve water supply and flood control operations at two projects in California by evaluating if they can retain storm water for future supply as long as the retained water can safely be released, if necessary, prior to the next storm. Knowledgeable stakeholders GAO interviewed said it is important for the Corps to consider forecast-based operations at its projects to help ensure efficient operations and to be able to respond to changing patterns of precipitation. Corps officials said the agency may consider doing so once the two California projects are completed in 2017.

* * *

July 26, 2016

The Honorable James Inhofe
Chairman

The Honorable Barbara Boxer
Ranking Member

Committee on Environment and Public Works
United States Senate

The Honorable Bill Shuster
Chairman

The Honorable Peter DeFazio
Ranking Member
Committee on Transportation and Infrastructure
House of Representatives

The U.S. Army Corps of Engineers (Corps) is the world's largest public engineering agency, with water resources projects across the United States, including more than 700 dams that it owns and operates for a variety of purposes including navigation, flood control, irrigation, hydropower, water supply, recreation, and fish and wildlife conservation. However, much of the Corps' infrastructure for these dams and their associated reservoirs was built more than 50 years ago. To manage and operate each water resources project, the Corps' 38 district offices develop water control manuals to guide project operations. These manuals describe the project's dams, reservoirs, and any affected rivers; historic floods and storms in the project area; and data from other agencies, such as the Department of Commerce's National Oceanic and Atmospheric Administration (NOAA) and the Department of the Interior's U.S. Geological Survey (USGS), that the Corps uses in operating the projects. The manuals also, among other things, describe methods for forecasting the amount of runoff flowing to the dams' reservoirs,[1] document policies and procedures for deciding how much water to release from the reservoirs, and generally have an associated drought contingency plan that provides guidance for district actions in response to periods of water shortages. Stakeholders, such as local governments and advocacy groups, have raised concerns that some water control manuals have not been revised since projects were built decades ago and may not reflect advances in science, such as in weather forecasting or changes in weather patterns.

In addition, the Corps' projects may be affected by extreme weather events, such as flood and drought. According to the U.S. Global Change Research Program's May 2014 National Climate Assessment and a 2010 National Research Council's report, precipitation patterns are changing, and the frequency and intensity of some extreme weather events are increasing.[2] In addition, a 2009 USGS report found that changes in the climate could affect water resources management and require changed operational assumptions about resource supplies, system demands or performance requirements, and operational constraints.[3] For example, according to the report, a shift in precipitation from snow to rain, combined with earlier melting of mountain snowpack, has been documented in western states. These two reported actions change the timing of runoff, affecting the Corps' operational decisions about when to release water from reservoirs.

Section 1046 of the Water Resources Reform and Development Act of 2014 includes a provision for us to audit the Corps' reviews of project operations, including an assessment of whether the Corps' practices could result in greater efficiencies to better prepare for extreme weather. According to a 1992 Corps' report,[4] project operations are defined as water control management that is routinely required to control either water level or flow, or both. Our report (1) examines the extent to which the Corps has reviewed or revised selected water control manuals and (2) describes the Corps' efforts, if any, to improve its ability to respond to extreme weather. Section 1046 of the Water Resources Reform and Development Act of 2014 also required the Corps to update its 1992 report about authorized and operating purposes of its reservoirs by June 10, 2016. The updated report is, among other things, to include a plan for reviewing the operations of individual projects that meet specified requirements.[5] The act also includes a provision for us to review the plan in the updated report. However, according to officials from the Corps and from the office of the Assistant Secretary of the Army for Civil Works, the Corps did not update the report as required by the statutory deadline because of funding constraints; therefore, we were unable to review the updated report.

To address both objectives, we reviewed relevant laws and executive orders, as well as our past reports on Corps operations and preparation for extreme weather. We interviewed officials who are responsible for project operations at headquarters and all eight Corps divisions. In addition, we interviewed officials from a nonprobability sample of 15 of the 38 Corps districts responsible for water control management.[6] We selected these districts based on criteria such as the Corps division where the district resides, regional differences in weather conditions, and a range in the number of projects operating within the district. Because this was a nonprobability sample, our findings cannot be generalized to all Corps districts but provide illustrative examples of the Corps' actions and strategies. We also visited two district offices (Los Angeles and Sacramento Districts) from our sample, where we reviewed documents, interviewed agency officials, and observed some of the Corps' efforts to help prepare its operations for extreme weather.

We selected these offices based on available resources and proximity to a GAO field office, in addition to the criteria used to select the nonprobability sample of districts. The results from these visits are also not generalizable.

To examine the extent to which the Corps has reviewed and revised selected water control manuals, we reviewed relevant Corps guidance and other documents, and information on water control manual reviews or revisions, such as engineer regulations, circulars, and manuals. We compared agency practices with federal standards for internal control.[7] We focused on steps taken since 1990 because of enacted legislation that changed the Corps' process to revise water control manuals and because 1990 was the last time the Corps systematically prepared drought contingency plans for water control manuals.[8] In addition, we examined the Corps' relevant documents and information on water control manual reviews and revisions from each of the eight divisions and our nongeneralizable sample of districts.

To examine the Corps' efforts, if any, for improving its ability to respond to extreme weather, we examined the Corps' relevant guidance, documents, and information on project operations and efforts to prepare for or respond to extreme weather. We interviewed cognizant Corps officials to discuss what, if any, guidance is being drafted or implemented responding to extreme weather events at the Corps' projects. We also interviewed five knowledgeable stakeholders from academia, a consulting firm, and a federal science agency to obtain their views on leading practices in preparing operations for extreme weather. Knowledgeable stakeholders were selected based on their knowledge of reservoir operations, modeling precipitation patterns, and Corps operational decisions. Views from these knowledgeable stakeholders are not generalizable to those with whom we did not speak.

We conducted this performance audit from August 2015 to July 2016 in accordance with generally accepted government auditing standards. Those standards require that we plan and perform the audit to obtain sufficient, appropriate evidence to provide a reasonable basis for our findings and conclusions based on our audit objectives. We believe that the evidence obtained provides a reasonable basis for our findings and conclusions based on our audit objectives.

BACKGROUND

This section provides information on the Corps' organizational structure, its project operations and water control manuals, and the process for formulating its operations and maintenance budget.

Corps' Organizational Structure

Located within the Department of Defense, the Corps has both military and civilian responsibilities.[9] The Corps' civil works program is organized into three tiers: a national headquarters in Washington, D.C.; eight regional divisions that were established generally according to watershed boundaries; and 38 districts nationwide (see fig. 1).

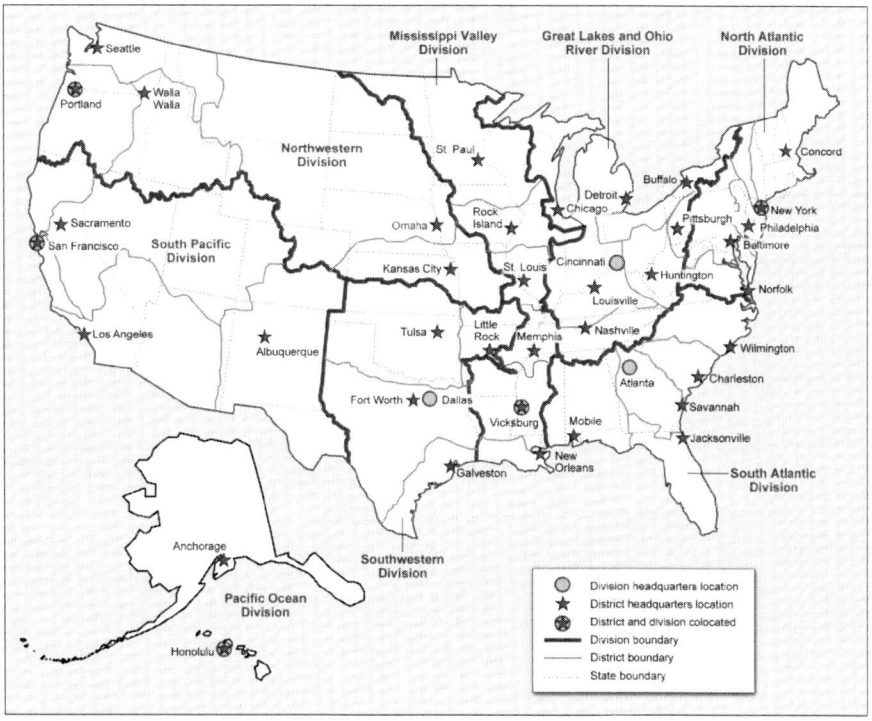

Figure 1. Locations of the U.S. Army Corps of Engineers' Civil Works Divisions and Districts.

Corps headquarters primarily develops policies and provides oversight. The Assistant Secretary of the Army for Civil Works, appointed by the President, establishes the policy direction for the civil works program. The Chief of Engineers, a military officer, oversees the Corps' civil works operations and reports on civil works matters to the Assistant Secretary of the Army for Civil Works. The eight divisions, commanded by military officers, coordinate civil works projects in the districts within their respective divisions. Corps districts, also commanded by military officers, are responsible for planning, engineering, constructing, and managing water-resources infrastructure projects in their districts. Districts are also responsible for coordinating with projects' nonfederal sponsors, which may be state, tribal, county, or local governments or agencies.[10]

In 1969, the Corps formed the Institute for Water Resources—which is a field-operating activity outside of the headquarters, division, and district structure—to provide forward-looking analysis and research in developing planning methodologies to aid the civil works program. Specifically, the institute fulfills its mission, in part, by providing an analysis of emerging water resources trends and issues and state-of-the-art planning and hydrologic-engineering methods, models, and training. In 2009, the Corps established the Responses to Climate Change program under the lead of Institute for Water Resources to develop and implement practical, nationally consistent, and cost-effective approaches and policies to reduce potential vulnerabilities to the nation's water infrastructure resulting from climate change and variability.

Project Operations and Water Control Manuals

The Corps is responsible for operations at 707 dams that it owns at 557 projects across the country, as well as flood control operations at 134 dams constructed or operated by other federal, nonfederal, or private agencies. Each of these projects may have a single authorized purpose or serve multiple purposes such as those identified in the original project authorization, revisions within the discretionary authority of the Chief of Engineers, or project modifications permitted under laws enacted subsequent to the original authorization. For example, the Blackwater Dam in New Hampshire has the single purpose of flood control, whereas the Libby Dam in Montana has multiple purposes, including hydropower, flood control, and recreation.

These 841 dams and their reservoirs are operated according to water control manuals and their associated water control plans, which Corps regulations require to be developed.[11] A water control manual may outline operations for a single project or a system of projects. For example, the Missouri River Mainstem Reservoir System Master Water Control Manual outlines the operations at six dams and their associated reservoirs, and the Folsom Dam Water Control Manual applies to one dam and its reservoir. Water control manuals include a variety of information the Corps uses in operating the dams, including protocols for coordinating with and collecting data from federal agencies, such as NOAA's National Weather Service and USGS, as well as water control plans. The water control plans, sometimes referred to as chapter 7 of the water control manuals, outline how each reservoir is to be operated and include relevant criteria, guidelines, and rule curves defining the seasonal and monthly limits of storage and guide water storage and releases at a project. According to the Corps' engineer regulations, the Corps develops water control plans to ensure that project operations conform to objectives and specific provisions of authorizing legislation. Water control plans also generally describe how a reservoir will be managed, including how water is to be allocated between a flood control storage pool and a conservation storage pool, which is used to meet project purposes during normal and drought conditions.[12] The bottom of a conservation storage pool is considered inactive and is designed for collecting sediment (see fig.2).

Water levels in the pools are defined based on a statistical analysis of historical rain events. For those projects that have multiple authorized purposes, water control plans attempt to balance water storage for all purposes.

Corps engineer regulations require that all water control manuals—except manuals for dry reservoirs that do not fill with water unless floodwaters must be contained—have an associated drought contingency plan to provide guidance for water management decisions and responses to a water shortage due to climatological drought.[13] These plans, which can cover more than one project: (1) outline the process for identifying and monitoring drought at a project, (2) inform decisions taken to mitigate drought effects, and (3) define the coordination needed with stakeholders and local interests to help manage water resources so they are used in a manner consistent with the needs that develop, among other things.

The conservation storage pool may be used for hydropower generation, water supply, recreation, and navigation, among other uses, and the inactive storage pool collects sediment.

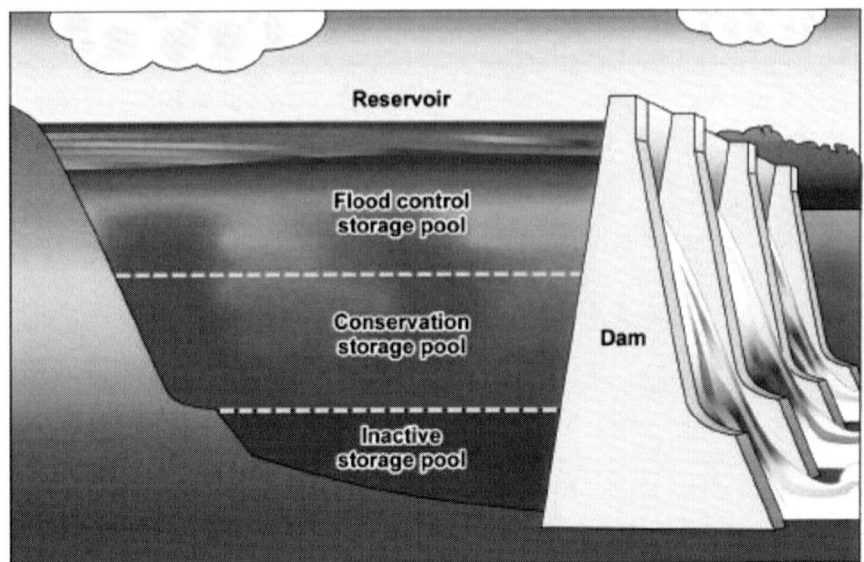

Source: GAO. | GAO-16-685.

Figure 2. Illustration of a U.S. Army Corps of Engineers' Reservoir, with Pools Allocated for Different Uses.

According to a 2014 Corps engineer regulation,[14] water control manuals may be revised for reasons such as land use development in the project area and downstream from it, improvements in technology used to operate projects, reallocation of the water supply,[15] new regional priorities, or changes in environmental conditions. The Corps' engineer regulation also directs districts to include in water control manuals a provision allowing temporary deviations from a project's approved water control plan to alleviate critical situations, such as a flood or drought, or to realize additional efficiencies without significantly affecting the project's authorized purposes. Districts are to perform a risk and uncertainty analysis to determine the potential consequences of such a deviation. Division commanders are responsible for reviewing and approving any proposed deviations. According to the engineer regulation, deviations are meant to be temporary and, if a deviation lasts longer than 3 years, the water control manual must be revised.

Corps' Operations and Maintenance Budget Formulation Process

Our prior work has found that the Corps' headquarters, divisions, and districts are all involved in developing the President's budget request for the Corps.[16] The development process spans 2 years; for example, development of the fiscal year 2018 budget began in fiscal year 2016. After receiving budget guidance from the Office of Management and Budget as well as the Assistant Secretary of the Army for Civil Works, district staff compile a list of operations and maintenance (O&M) projects necessary in their districts and submit their needs to the relevant division. O&M projects may include, among other things, water control manual revisions, dredging, replacement of dam parts, dam safety measures, or adding capacity at hydropower projects. Division staff then rank the O&M projects from all districts in the division and submit those rankings to Corps headquarters staff for review. Headquarters staff review the rankings to help ensure they are consistent with Corps-wide guidance and result in decisions that emphasize agency-wide priorities. Headquarters staff consolidate the O&M requests across business lines and divisions into a highest-priority grouping. Once the Corps completes its internal review of the budget request, the Assistant Secretary of the Army for Civil Works approves and submits its budget to the Office of Management and Budget for review. The Office of Management and Budget recommends to the President whether to support or change the Corps' budget request, and the President's budget request is transmitted to Congress.

THE EXTENT TO WHICH THE CORPS HAS REVIEWED OR REVISED WATER CONTROL MANUALS IS UNCLEAR BECAUSE IT DID NOT DOCUMENT REVIEWS OR TRACK REVISIONS

According to agency officials, the Corps conducts ongoing, informal reviews of selected water control manuals and has revised some of them, but the extent of the reviews and revisions is unclear because they were not documented or tracked. More specifically, district officials said that the Corps reviews the manuals as part of daily operations but does not document the reviews, and there is no guidance on what constitutes a review or how to document it. Further, the Corps does not track consistent information across divisions on the status of manuals to indicate revisions that were made or are needed.

The Extent to Which the Corps Has Reviewed Its Water Control Manuals Is Unclear Because the Reviews Were Informal, According to Agency Officials, and Not Documented

It is unclear to what extent the Corps has reviewed its water control manuals because district officials did not document these reviews, which, according to district officials, are informal and conducted on an ongoing basis through daily operations. A 2014 Corps engineer regulation states that water control manuals should be reviewed no less than every 10 years, so that they can be revised as necessary.[17] Most district officials we interviewed said that they informally review the water control plan because this portion of the manual describes how projects are to be operated under different conditions to meet their authorized purposes. However, officials we interviewed from all 15 districts said they do not document these informal reviews because they consider such reviews to be part of the daily routine of operating projects. Because these informal reviews are not documented, knowledge of these reviews and their results may be limited to personnel directly involved with them. Officials we interviewed from four districts said that the loss of institutional knowledge posed a challenge to conducting efficient reviews of manuals. For example, officials from one district said that no Corps officials currently employed at the district had worked on developing the manual for a project and had no supporting documentation of the process, so the officials did not know why prior Corps officials wrote the manual in a particular way. As a result, the officials said it took them longer to review the manual.

One Corps district we reviewed had previously documented informal reviews of water control manuals. Specifically, officials we interviewed in this district said that they documented reviews of some water control manuals in 2005 as part of a district-wide effort to ensure these manuals were adequate to meet the projects' authorized purposes since they had not been revised in a long time. According to these officials, as part of this effort, if they determined that all of the operating conditions in a manual were still current, they submitted a memorandum to their division that revalidated the manual's water control plan. Officials from that district said they have not documented reviews of water control manuals since 2005 because they chose to focus only on those manuals they knew needed revision. However, the Corps does not have guidance on what activities constitute a review or how officials should document the results of their reviews. Under federal standards for internal control, internal control and other significant events are to be clearly documented in a manner that allows the documentation to be readily available for examination, such as in management directives, administrative policies, or operating manuals.[18] Without developing guidance on what activities constitute a review of a water control manual and how to document that review, the Corps does not have reasonable assurance that its districts will consistently conduct reviews and document them to provide a means to retain organizational knowledge and mitigate the risk of having that knowledge limited to personnel directly involved with these reviews.

The Corps Has Revised Some Water Control Manuals but Does Not Track Consistent Information on Which Have Been Revised or Need Revision

The Corps has revised some water control manuals; however, divisions and districts do not track consistent information about revisions to manuals, and the extent to which they have been revised—or need revision—is unclear. Corps engineer regulations state that manuals are to be revised as needed, in accordance with the regulations.[19] Districts have revised some water control manuals for a variety of reasons, such as in response to infrastructure modifications and weather events, according to the Corps' documents and its headquarters, division, and district officials we interviewed. For example, officials we interviewed in one district said they revised a water control manual after a flood highlighted a need to change the seasonal and monthly limits of reservoir storage when water recedes. Officials we interviewed from other districts said they revised a manual based on vulnerabilities identified through the periodic inspections they conduct of projects through the Corps' dam safety program.[20]

District officials we interviewed said that the time and resources needed to revise manuals vary greatly, depending on the nature of the revisions and the complexity of the project, among other things. For instance, according to a Corps 2012 engineer regulation,[21] all revisions to a water control manual are to undergo a quality control review of the science and engineering work by district leadership. Depending on the revisions made, manuals may also undergo a technical review by division leadership and an independent external peer review by a panel of experts. For example, according to a Corps engineer regulation and division and district officials we interviewed,[22] if the districts make substantial revisions to a manual's water control plan, they are to complete environmental analyses required by the National Environmental Policy Act of 1969,[23] which they said involves considerable time and coordination with other federal agencies and opportunity for public comment. District officials told us that making such substantive revisions to a manual takes more time and resources than making an administrative revision because of the additional requirements for review.[24]

Moreover, some district officials noted that the longer they defer making revisions to a manual, the more extensive and complex the changes may become, changes that may add time and increase costs to revise the manual. Officials in one district said that it cost about $100,000 to revise one section of a manual's water control plan, which did not significantly affect other aspects of the plan. In contrast, officials in another district said that it cost over $10 million and took over 25 years to revise a manual that included a water control plan for several projects, primarily because of litigation over the revisions.

Our review of division documents indicates that all eight divisions we reviewed tracked the date a manual was last revised, but officials told us that the length of time since the last revision is not necessarily indicative of whether manuals need to be revised. According to headquarters and district officials we interviewed, water control manuals are designed to provide flexibility for a broad variety of runoff and climatic conditions. For example, headquarters officials said the rule curve[25] in one water control manual provided guidelines for how much water operators should take out of the reservoir during October and November to meet its flood risk management target, while at the same time holding enough water to, among other things, meet its authorized purposes of hydropower and providing water flow for an endangered fish species. However, two knowledgeable stakeholders we interviewed said that many of the Corps' rule curves assume that the chances of an extreme event are equally likely for any given year, which may not reflect actual conditions. These stakeholders said that the Corps should consider revising water control manuals with dynamic rule curves to account for potential changes to climate conditions,[26] but a Corps official said that the science behind dynamic rule curves is still being developed.

In addition, Corps officials said that the provisions in water control manuals that allow temporary deviations from water control plans, if necessary, provide districts with flexibility in operating projects. For example, in response to drought conditions, the Corps approved a deviation from the water control plan in December 2014 at a project in California, a deviation that allowed the Corps to temporarily retain water captured behind the dam following a rainstorm. According to officials in that district, this temporary deviation allowed them to respond to the immediate stakeholder interests in conserving water during the drought, so they did not need to revise the water control manual. Given the flexibilities provided by rule curves and temporary deviations, not all manuals need to be revised, according to Corps officials we interviewed at headquarters, divisions, and districts.

However, the extent to which water control manuals have been substantively revised, if at all, remains unknown because the divisions and districts we reviewed did not track consistent information about revisions to water control manuals to help ensure that manuals are revised in accordance with engineer regulations. For example, based on our review of Corps documents, one of eight divisions tracked whether the water control plans in its water control manuals reflected actual operations of the project, but the remaining seven divisions did not. In addition, another division tracked information about when the water control manuals in five out of six of its districts had been revised. Officials whom we interviewed from this division said they were not sure if any of the manuals in the sixth district had been reviewed because information had not been submitted by the district. Corps headquarters officials said that the Corps does not track the status of water control manual revisions agency-wide because two people in headquarters oversee all of the Corps' water resources operational issues, among other duties, and, therefore, divisions and districts were given responsibility for tracking revisions. However, these officials said the agency is compiling information to create a central repository of water control manuals, among other things, to respond to activities set forth in an action plan for the President's Memorandum on drought resilience.[27] They said the repository could be used to track the status of revisions or needed revisions of manuals, but they do not currently plan to do so.

Furthermore, district officials we interviewed told us they have identified certain manuals needing revision, but they have not received the O&M funds they requested to revise these manuals and documentation shows that they do not track consistent information on these manuals. A Corps engineer manual states that there may be reasons—such as new hydrologic data or a reevaluation of water control requirements—to revise water control manuals to reflect current operating conditions.[28] Divisions are responsible for prioritizing the O&M funding requests they receive from all of their districts. Corps budget documents describe factors to consider for agency-wide prioritization—such as whether an item is required to meet legal mandates or would help ensure project safety (e.g., by paving a project access road)—but headquarters officials said each division may add other factors for consideration. According to our document review, one of the eight divisions tracked the priority that districts assigned to revising water control manuals when requesting O&M funds during the budgeting process, and four divisions tracked the fiscal year they proposed revising certain manuals, pending available funding. However, most district officials we interviewed said revisions to water control manuals are often a lower priority than other O&M activities, such as equipment repairs, sediment removal, or levee repairs. As a result, districts may not get funding to revise water control manuals.

Moreover, Corps headquarters officials said that each division and district varies in the resources and staff it has available to conduct water control manual reviews and make revisions. For example, officials we interviewed from two districts in the same division said they do not have staff available to review water control manuals, and they have not received the funding they requested to revise their water control manuals. Corps headquarters officials said they do not track which manuals the districts have requested funds to revise—and therefore cannot prioritize these requests—because they have limited staff to accomplish water resources management activities. However, internal control standards in the federal government call for agencies to clearly and promptly document transactions and other significant events from authorization to completion.[29] Without tracking which manuals need revision, it is difficult for the Corps to know the universe of projects that may not be operating in a way that reflects current conditions as called for in the Corps' engineer manual and prioritize revisions as needed.

District officials whom we interviewed said that not revising water control manuals regularly could lead projects to operate inefficiently under changing conditions. For example, farmers downstream from one project wanted the Corps to consider changing operations so that their fields would not flood when it rained. However, officials in that district said they requested but did not receive the funds to revise the manual and could not fully address the farmers' concerns. Officials in another district said they have requested funds to revise several manuals that they described as outdated, but because they have not received funds, they noted they were operating those projects in a way that differed from some aspects of the approved water control plans and they did not request deviations. Instead, they said they referred to handwritten notes and institutional knowledge to operate those projects. For example, officials said that due to sedimentation build up in the reservoir of one project, they are operating that project 22 feet higher than the approved plan.

According to a Corps engineer regulation, the Corps develops water control plans to ensure that project operations conform to objectives and specific provisions of authorizing legislation.[30] However, because some manuals that need revision have not been revised and, as some district officials noted, operations for certain projects differ from aspects of the approved water control plans in those manuals, the Corps lacks assurance that project operations are conforming to the objectives and specific provisions of authorizing legislation.

THE CORPS HAS EFFORTS UNDER WAY TO IMPROVE ITS ABILITY TO RESPOND TO EXTREME WEATHER

The Corps has efforts under way to improve its ability to help respond to extreme weather events. These efforts include developing a strategy to revise its drought contingency plans and studying the use of forecasts to make decisions on project operations. The Corps is also conducting research on how to better prepare operations for extreme weather.

The Corps Is Developing a Strategy to Revise Drought Contingency Plans and Is Studying the Use of Forecasts to Make Operations Decisions at Two Projects

To better respond to drought, the Corps is developing a strategy to analyze drought contingency plans in its manuals and devise methods for those plans to account for a changing climate. According to a 2015 Corps report on drought contingency planning, the Corps is developing the strategy because climate change has been and is anticipated to continue to affect the frequency and duration of drought in the United States.[31] The Corps last systematically prepared drought contingency plans in the 1980s through the early 1990s, before climate change information was widely available. These plans assumed that historic patterns of temperature, precipitation, and drought provided a reasonably accurate model of future conditions. According to the Corps' 2015 report, the agency subsequently identified and reviewed all of its drought contingency plans. The Corps' review found (1) that none of the plans contained information on drought projections under future climate change and

(2) that it was unlikely that the plans provided an adequate guide for preparing for future droughts.

As of May 2016, the Corps was conducting pilot updates of drought contingency plans at five high-priority projects to help test methods and tools for those plans to account for a changing climate. According to the Corps' 2015 report, these pilot projects will help the agency develop a framework for a systematic update of drought contingency plans. Corps officials said these pilots are to be largely completed by the end of calendar year 2016. The Corps has created an internal website available to all Corps officials to disseminate the results of the drought contingency plan analysis, pilot project results, and other drought-related information. In addition to completing the pilot projects, Corps officials said the agency plans to compile a list of drought contingency plan priorities by the middle of fiscal year 2017 for inclusion in the fiscal year 2018 budget.

In addition to its efforts related to drought contingency plans, the Corps is studying the use of forecasting tools to determine whether water control manuals can be adjusted to improve water-supply and flood-control operations at two projects in California—Folsom Dam and Lake Mendocino.[32] The Corps has historically used forecasts to some degree in its operations, largely by using models that create a single forecast based on the existing hydrologic data. According to Corps officials, the Folsom Dam and Lake Mendocino projects are evaluating the potential to incorporate forecasts into their operational rules, by using statistical techniques to simulate multiple, slightly different initial conditions and identify a range of potential outcomes and their probability. The use of forecasts at these projects will depend on whether the skill of the forecasts is improved to the point where they are viable in informing reservoir operations.[33] Corps officials told us that the forecasts must be accurate in terms of space and time to allow the reservoirs to retain some water for future supply as long as the retained water can be safely released, if necessary, prior to the next storm.

At the first project, Folsom Dam, the Corps and the Department of the Interior's Bureau of Reclamation are constructing an auxiliary spillway project to improve the safety of the dam and reduce the flood risk for the Sacramento area.[34] Officials also said the water control manual must be updated to reflect the physical changes to the project, but the Corps is also considering incorporating forecasting into its operating rules so that prior to storm events, water can be released earlier than without forecasting capabilities. Corps officials said the revisions to the Folsom Dam water control manual, outlining the forecast-based operations, are estimated to be completed in April 2017. For the second project, Lake Mendocino, an interagency steering committee was formed to explore methods for better balancing water supply needs and flood control by using modern forecasting observation and prediction technology. Corps officials told us the interagency committee expects to complete a preliminary viability study on the project by the end of calendar year 2017.

Corps headquarters officials said that once they determine how forecasting can be incorporated into these projects, the agency may consider using forecast-based operations at other projects. Four of the five knowledgeable stakeholders we interviewed said that it would be important for the Corps to consider using such operations to help ensure efficiency and to be able to respond to changing patterns of precipitation. These views are consistent with our 2014 report on the Missouri River flood and drought of 2011 to 2013, in which we recommended that the Corps evaluate forecasting techniques that could improve its ability to anticipate weather developments for certain projects.[35] However, Corps officials and knowledgeable stakeholders also said that the Corps faces two key challenges in implementing forecast-based operations at its reservoirs. First, four of the five knowledgeable stakeholders we interviewed said that the Corps' primary mission of flood control makes it difficult for the agency to accept the uncertainty that is involved with forecasting. Second, forecasting may be more complex in certain regions of the country, because according to one knowledgeable stakeholder and Corps officials, much of the rain in California is a result of atmospheric rivers,[36] which produce rainfall that is more predictable than the convection rains that are experienced in the Midwest.[37]

The Corps Is Conducting Research on How to Better Prepare Operations for Extreme Weather

The Corps' Responses to Climate Change program is conducting research on adaptation measures through vulnerability assessments for inland projects and sedimentation surveys.[38] In 2012, the Corps initiated an initial vulnerability assessment that focused on how hydrologic changes due to climate change may impact freshwater runoff in some watersheds. This assessment identified the top 20 percent of watersheds most vulnerable to climate change for each of the Corps' business lines.[39] According to Corps officials, this assessment was conducted for watersheds, because actionable science was not currently available to conduct such an assessment at the project level. However, the Corps is working with an expert consortium of federal and academic organizations—including NOAA, the Bureau of Reclamation, USGS, the University of Washington, and the University of Alaska—to develop future projected climatology and hydrology at finer scales. This project is intended to provide the Corps and its partners and stakeholders with a consistent, 50-state strategy to further assess vulnerabilities, a strategy that will also support planning and evaluation of different adaptation measures to increase resilience to specific climate threats. According to the Corps, this consortium holds monthly meetings to review progress made by the various members. According to Corps officials, the consortium plans to release reports in 2016 and 2017 that will enable the Corps to improve tools, methods, and guidance for finer-resolution analyses using climate-impacted hydrology.

The Corps has also begun to evaluate reservoir vulnerabilities to altered sedimentation rates resulting from extreme weather and land use changes. In 2012, the Corps began conducting 15 pilot studies at various districts to test different methods and serve as a framework for adapting to climate change. Two of these pilots predicted changes in the amount of sediment in a reservoir because of changes in hydrologic variables as a result of climate change. Additionally, according to the Corps' website, reservoirs in areas with drought conditions have experienced lower-thannormal levels of water in their conservation storage pools. These lower levels have revealed additional and unexpected sedimentation in reservoirs that could reduce the space available to store water. In 2013, the Corps developed a program to deploy airborne laser scanning systems to measure and collect data on the reservoirs in drought-affected areas. In 2015, this system was tested in California to refine the process to collect sedimentation data and modify the system for specific aircraft. According to a Corps official we interviewed in the Responses to Climate Change program, the agency plans to further refine the data collected and evaluate how these data change over time. This effort, the official told us, is also expected to provide indicators to support the analysis of future sedimentation rates based on climate changes for use in the Corps' climate vulnerability analysis. The official said a baseline report on the Corps' reservoir sedimentation status is expected by the end of fiscal year 2016. This effort was highlighted in the action plan for the President's Memorandum on Building National Capabilities for Long-Term Drought Resilience, which lays out a series of activities to fulfill the President's drought-resilience goals.[40]

CONCLUSION

The Corps has revised some of the water control manuals used to operate its water resources projects, which serve important public purposes such as flood control, irrigation, and water supply. But district officials told us there are manuals that do not reflect the changing conditions in the areas surrounding the projects. A Corps engineer regulation states that the water control manuals should be reviewed no less than every 10 years and revised as needed. However, there is no Corps guidance on what activities constitute a review, and while district officials said they informally reviewed selected water control manuals through daily operations, they also said they do not document these reviews. Without developing guidance on what activities constitute a review of a water control manual and how to document that review, the Corps does not have reasonable assurance that its districts will consistently conduct reviews and document them to provide a means to retain organizational knowledge and mitigate the risk of having that knowledge limited to personnel directly involved with these reviews.

In addition, while the Corps has revised certain water control manuals in accordance with its engineer regulation, it does not track consistent information on revisions to its manuals. Furthermore, district officials said that they have requested funds to revise additional water control manuals as needed to reflect changing conditions, but they have not received those funds, and have not tracked consistent information about manuals needing revisions. However, internal control standards in the federal government call for agencies to clearly and promptly document transactions and other significant events from authorization to completion. Without tracking which manuals need revision, it is difficult for the Corps to know the universe of projects that may not be operating in a way that reflects current conditions as called for in the Corps' engineer manual and to prioritize revisions as needed. Because some manuals that need revision have not been revised and some district officials noted that operations for certain projects differ from aspects of the approved water control plans in those manuals, the Corps lacks assurance that project operations are conforming to the objectives of authorizing legislation.

RECOMMENDATIONS FOR EXECUTIVE ACTION

To help improve the efficiency of Corps operations at reservoir projects and to assist the Corps in meeting the requirement of the Water Resources Reform and Development Act of 2014 to update the Corps' 1992 reservoir report, we recommend that the Secretary of Defense direct the Secretary of the Army to direct the Chief of Engineers and Commanding General of the U.S. Army Corps of Engineers to take the following two actions:

- develop guidance on what activities constitute a review of a water control manual and how to document that review; and
- track consistent information on the status of water control manuals, including whether they need revisions, and prioritize revisions as needed.

AGENCY COMMENTS

We provided a draft of this report for review and comment to the Department of Defense. In its written comments, the department concurred with our recommendations and noted that it will take steps to address these recommendations as it updates its guidance. In its comments, the department also stated that, as of May 2016, it had updated its Engineer Regulation 1110-2-240, *Engineering and Design: Water Control Management*. We incorporated this information into the report.

Anne-Marie Fennell
Director,
Natural Resources and Environment

End Notes

[1] Runoff flows over the land surface, going downhill into rivers and streams.
[2] Jerry M. Melillo, Terese (T.C.) Richmond, and Gary W. Yohe, eds., 2014: Climate Change Impacts in the United States: The Third National Climate Assessment, (Washington, D.C.: U.S. Global Change Research Program, May 2014) and National Research Council, Board on Atmospheric Sciences and Climate, Advancing the Science of Climate Change (Washington, D.C.: National Academies Press, 2010). The U.S. Global Change Research Program coordinates and integrates the activities of 13 federal agencies that conduct

research on changes in the global environment and their implications for society. The National Research Council is now known as the National Academies of Sciences, Engineering, and Medicine.

[3] L.D. Brekke, J.E. Kiang, J.R. Olsen, R.S. Pulwarty, D.A. Raff, D.P. Turnipseed, R.S. Webb, and K.D. White, Climate Change and Water Resources Management—A Federal Perspective: USGS Circular 1331 (2009).

[4] Army Corps of Engineers, Authorized and Operating Purposes of Corps of Engineers Reservoirs (Washington, D.C.: July 1992). This report excludes water control structures that do not routinely impound water, such as river diversion structures and pumping stations.

[5] Pub. L. No. 113-121, § 1046(a)(2)(B)(ii)(I)(dd) (2014).

[6] We interviewed Corps officials from the following districts: Anchorage, Alaska; Philadelphia, Pennsylvania; Jacksonville, Florida; Walla Walla, Washington; St. Louis, Missouri; Los Angeles, California; Sacramento, California; Kansas City, Missouri; Fort Worth, Texas; Concord, Massachusetts; St. Paul, Minnesota; Louisville, Kentucky; Tulsa, Oklahoma; Pittsburgh, Pennsylvania; and Mobile, Alabama.

[7] GAO, Standards for Internal Control in the Federal Government, GAO/AIMD-00-21.3.1 (Washington, D.C.: Nov. 1, 1999). GAO has revised and reissued Standards for Internal Control in the Federal Government, with the new revision effective as of October 1, 2015. GAO-14-704G (Washington, D.C.: September 2014).

[8] Pub. L. No. 101-640, tit. III, § 310 (1990) (codified as amended at 33 U.S.C. § 2319).

[9] The Corps' military program provides, among other things, engineering and construction services to other U.S. government agencies and to foreign governments. This report focuses only on the Corps' civil works program.

[10] Nonfederal sponsors are those entities that share the cost of planning and implementing Corps projects. The division of federal and nonfederal cost-sharing required varies by project purpose.

[11] 33 C.F.R. § 222.5(f)(1).

[12] The flood storage pool captures intense rainfall and provides flood risk reduction for downstream areas. The conservation storage pool may be used for hydropower generation, water supply, recreation, fish and wildlife habitat enhancement, and navigation, among other uses.

[13] U.S. Army Corps of Engineers, Engineering and Design: Water Control Management, Engineer Regulation 1110-2-240 (Washington, D.C.: June 30, 2014) and Engineering and Design: Drought Contingency Plans, Engineer Regulation 1110-2-1941 (Washington, D.C.: Sept. 15, 1981). The Corps updated Engineer Regulation 1110-2-240 on May 30, 2016.

[14] Engineer Regulation 1110-2-240.

[15] Reallocation of water supply storage occurs when the storage is changed from one authorized purpose to a different authorized use. For example, to meet increased demand for water by cities, water storage within a reservoir may be reallocated from hydropower to municipal and industrial use.

[16] GAO, Army Corps of Engineers: Budget Formulation Process Emphasizes Agencywide Priorities, but Transparency of Budget Presentation Could Be Improved, GAO-10-453 (Washington, D.C.: Apr. 2, 2010).

[17] Engineer Regulation 1110-2-240.

[18] GAO/AIMD-00-21.3.1.

[19] Engineer Regulation 1110-2-240; U.S. Army Corps of Engineers, Engineering and Design: Management of Water Control Systems, Engineer Manual 1110-2-3600 (Washington, D.C.: Nov. 30, 1987).

[20] The Corps is required by law to carry out a national program of inspection of dams for the purpose of protecting human life and property. 33 U.S.C. § 467a. Through this program, the Corps conducts: (1) annual inspections to ensure that a dam is being properly operated and maintained; (2) periodic inspections every 5 years, including a more detailed, comprehensive evaluation of the condition of the dam; and (3) risk assessments every 10 years, including the probability of failure and resulting potential consequences due to failure.

[21] U.S. Army Corps of Engineers, Water Resources Policies and Authorities: Civil Works Review, Engineer Circular 1165-2-214 (Washington, D.C.: Dec. 15, 2012).

[22] Engineer Regulation 1110-2-240.

[23] Pub. L. No. 91-190, 83 Stat. 852 (1970) (codified as amended at 42 U.S.C. §§ 4321- 4347).

[24] Revisions to water control manuals can be administrative, such as updating contact information, or substantive, which would change the water control plan of the project, according to Corps documents.

[25] Rule curves define the seasonal and monthly limits of water storage and guide water storage and releases at a project.

[26] Dynamic rule curves change based on the present state of a system, such as storage levels, current inflow, or forecasted conditions.

[27] The White House, Long-Term Drought Resilience: Federal Action Plan of the National Drought Resilience Partnership (Washington, D.C.: March 2016). The President's Memorandum on Building National Capabilities for Long-Term Drought Resilience establishes federal drought resilience goals, among other things.

[28] Engineer Manual 1110-2-3600.

[29] GAO/AIMD-00-21.3.1.

[30] Engineer Regulation 1110-2-240.

[31] U.S. Army Corps of Engineers, USACE Drought Contingency Planning in the Context of Climate Change, Civil Works Technical Report, CWTS 2015-15, (Washington, D.C.: Sept. 2015).

[32] The32The Corps is undertaking this effort at Folsom Dam because the Water Resources Development Act of 1999 requires the Secretary of the Army to update the Dam's flood management plan to reflect improved weather forecasts, among other things. Pub. L. No. 106-53, §101(a)(6)(E), 113 Stat. 269, 274 (1999). According to Corps officials, the Lake Mendocino effort began at the request of the County of Sonoma Corps is undertaking this effort at Folsom Dam because the Water Resources Development Act of 1999 requires the Secretary of the Army to update the Dam's flood management plan to reflect improved weather forecasts, among other things. Pub. L. No. 106-53, §101(a)(6)(E), 113 Stat. 269, 274 (1999). According to Corps officials, the Lake Mendocino effort began at the request of the County of Sonoma.

[33] Forecasting skill is the statistical evaluation of the accuracy of forecasts or the effectiveness of detection techniques.

[34] The Bureau of Reclamation has carried out its mission to manage, develop, and protect water and related resources in 17 western states since 1902. The agency has led or provided assistance in constructing most of the large dams and water diversion structures in the West for the purpose of developing water supplies for irrigation, as well as for other purposes, including hydroelectric power generation, municipal and industrial water supplies, recreation, flood control, and fish and wildlife enhancement. For more information, see GAO, Bureau of Reclamation: Availability of Information on Repayment of Water Project

Construction Costs Could Be Better Promoted, GAO-14-764 (Washington, D.C.: Sept. 8, 2014).

[35] GAO, Missouri River Flood and Drought: Experts Agree the Corps Took Appropriate Action, Given the Circumstances, but Should Examine New Forecasting Techniques, GAO-14-741 (Washington, D.C.: Sept. 12, 2014). The Department of Defense agreed with our recommendation to evaluate the pros and cons of forecasting techniques.

[36] Atmospheric rivers are relatively narrow regions in the atmosphere that are responsible for most of the horizontal transport of water vapor outside of the tropics.

[37] In meteorology, the term convection is used specifically to describe vertical transport of heat and moisture in the atmosphere, especially by updrafts and downdrafts in an unstable atmosphere.

[38] Vulnerability assessments identify, quantify, and prioritize the vulnerabilities in a system. Sedimentation surveys determine the amount of sediment that is in a reservoir.

[39] U.S. Army Corps of Engineers' business lines include flood risk reduction, navigation, ecosystem restoration, hydropower, recreation, regulatory, water supply, and emergency management.

[40] The White House, Long-Term Drought Resilience: Federal Action Plan of the National Drought Resilience Partnership, (Washington, D.C.: March 2016).

INDEX

A

adaptation, 13, 14, 30, 97
agencies, 3, 5, 6, 8, 11, 12, 13, 14, 15, 16, 20, 21, 22, 28, 29, 30, 33, 34, 36, 38, 40, 42, 43, 44, 45, 46, 47, 48, 49, 50, 59, 76, 78, 82, 86, 87, 92, 94, 99, 100, 101
air temperature, 12
Alaska, 20, 29, 98, 101
appropriations, 8, 24, 25, 46, 47
Appropriations Act, 8, 32
assessment, 23, 24, 25, 26, 27, 28, 32, 41, 83, 97
Assistant Secretary of the Army for Civil Works (ASA (CW)), 7, 10, 15, 51, 56, 62, 63, 64, 65, 83, 86, 89
audit, 6, 28, 31, 38, 55, 83, 84
authority(ies), 8, 32, 33, 36, 52, 53, 58, 61, 62, 63, 64, 65, 66, 70, 71, 72, 75, 77, 78, 86

B

barriers, vii, 3, 4, 22, 24, 25, 26, 56
benefits, 27, 29, 53, 54, 59, 71

C

chain of command, 62
challenges, 5, 17, 21, 69, 97
clarity, 64, 65, 71
classification, 23, 42, 44, 57
climate, 2, 5, 6, 11, 12, 13, 14, 16, 17, 19, 20, 21, 22, 23, 26, 28, 81, 82, 86, 92, 95, 96, 97, 98
climate change, 6, 13, 14, 15, 16, 17, 19, 21, 22, 26, 28, 86, 95, 97, 98
climate change issues, 21
coastal region, 32
coastal storm damage, vii, 4
collaboration, 16, 20
communication, 29, 40, 52, 55, 64, 68, 71, 72
community(ies), 8, 29, 41, 42, 47, 50, 54
compensation, 32
complement, 46
complexity, 91
conference, 8, 21, 50
Congress, 2, 3, 8, 9, 10, 26, 31, 34, 37, 41, 43, 47, 49, 50, 58, 77, 89
conservation, 82, 87, 98, 101
conserving, 93
Constitution, 32
construction, vii, 4, 8, 9, 10, 15, 16, 24, 29, 30, 32, 49, 50, 52, 53, 55, 56, 58, 59, 60, 61, 64, 65, 67, 69, 70, 74, 76, 78, 101
contingency, 18, 19, 81, 82, 84, 87, 95, 96
control measures, 66
coordination, 14, 50, 87, 92

Corps Water Management System (CWMS), 1, 19
cost, vii, 9, 10, 24, 29, 36, 40, 52, 53, 54, 55, 57, 58, 59, 60, 61, 62, 63, 64, 66, 67, 68, 69, 70, 71, 72, 73, 75, 76, 86, 92, 101
Council on Environmental Quality (CEQ), 1, 14, 15, 30
customers, 54, 60, 66

D

Dam Safety, v, 23, 24, 31, 51, 53, 56, 57, 60, 61, 62, 63, 64, 65, 66, 71, 72, 74, 75, 77, 78
Dam Safety Action Classification (DSAC), 51, 57, 58, 77
Dam Safety Officer (DSO), 51, 56, 58, 62
damages, 16, 32, 37
dams, vii, 2, 3, 4, 10, 21, 22, 23, 26, 31, 52, 53, 54, 55, 56, 57, 59, 60, 61, 62, 63, 64, 65, 66, 67, 68, 71, 72, 73, 76, 77, 78, 79, 82, 86, 102
data collection, 21
database, 45, 49
decision makers, 9
degradation, 53, 61
Department of Agriculture, 5, 29, 49
Department of Commerce, 4, 28, 49, 82
Department of Defense (DOD), 6, 21, 31, 36, 48, 51, 52, 55, 72, 76, 85, 100, 103
Department of Energy, 60
Department of Homeland Security, 36, 48
Department of the Interior, 5, 49, 64, 82, 96
Departments of Agriculture, 28
detection techniques, 102
deviation, 3, 19, 88, 92
direct measure, 17
directives, 91
disaster, 34, 36, 40, 47
disaster assistance, 41
disaster relief, 34, 36, 47
displacement, 76
distribution, 11, 30
draft, 2, 27, 28, 38, 41, 42, 43, 48, 49, 70, 72, 77, 100
drainage, 64
drawing, 69, 70
drought, 3, 4, 18, 19, 81, 82, 84, 87, 88, 92, 93, 95, 96, 97, 98, 102

E

economic damages, 37
ecosystem restoration, 32, 103
education, 42, 43, 45, 46, 47
emergency, 18, 32, 103
emergency management, 32, 103
engineering, vii, 4, 8, 9, 10, 16, 21, 29, 35, 37, 44, 45, 55, 56, 57, 58, 76, 82, 86, 91, 101
environment, 5, 10, 26, 101
environmental conditions, 88
Environmental Protection Agency (EPA), 29, 49, 67, 78
environmental resources, 25
erosion, 30, 32, 39, 53, 57, 61, 63, 65, 77
evidence, 6, 17, 38, 55, 84
executive orders, 5, 14, 15, 16, 30, 34, 37, 83
expertise, 8, 29, 43, 77
exposure, 25, 30
extreme weather events, vii, 2, 3, 4, 5, 6, 11, 15, 16, 18, 20, 21, 22, 26, 28, 82, 84, 95

F

farmers, 94
federal agency, 64
Federal Emergency Management Agency (FEMA), v, 22, 29, 32, 33, 34, 35, 36, 37, 38, 39, 40, 41, 42, 43, 44, 45, 46, 47, 48, 49, 50
federal funds, 15, 71

Index

federal government, 21, 36, 47, 59, 67, 71, 77, 94, 99, 101
federal law, 34, 37, 55
Federal Register, 77
Fifth Amendment, 32
financial, 34, 37, 46, 47, 55, 64, 67, 71
fiscal year, 2, 5, 10, 13, 19, 24, 31, 32, 46, 50, 55, 58, 62, 74, 75, 76, 89, 94, 96, 98
fish, 32, 82, 92, 101, 102
Fish and Wildlife Service, 49
flexibility, 19, 92
flood, vii, 4, 7, 10, 14, 15, 16, 17, 18, 20, 23, 25, 27, 30, 31, 32, 36, 38, 39, 40, 41, 42, 43, 49, 50, 54, 56, 61, 63, 76, 79, 81, 82, 86, 87, 88, 91, 92, 94, 96, 97, 99, 101, 102, 103
floodgates, vii, 4, 56
flooding, 18, 20, 23, 31, 33, 36, 38, 40, 42
floodwalls, vii, 3, 4, 21, 22, 24, 25, 26, 33, 36, 38, 40
forecasting, 11, 13, 20, 22, 81, 82, 96, 97, 103
freshwater, 28, 31, 97
funding, 2, 3, 8, 13, 14, 19, 20, 25, 26, 47, 48, 50, 62, 64, 67, 74, 78, 83, 93, 94
funds, 3, 8, 10, 15, 32, 47, 50, 58, 59, 68, 71, 93, 94, 99

G

GAO, 1, 2, 3, 5, 7, 12, 25, 27, 30, 31, 32, 33, 34, 35, 45, 49, 51, 52, 53, 57, 60, 62, 66, 73, 75, 77, 78, 79, 80, 81, 84, 88, 101, 102, 103
governments, 8, 12, 29, 36, 41, 42, 43, 76, 82, 86, 101
groundwater, 31
grouping, 89
guidance, 2, 3, 5, 6, 9, 10, 14, 15, 16, 17, 18, 19, 23, 52, 53, 64, 68, 71, 72, 78, 80, 82, 84, 87, 89, 91, 98, 99, 100
guidelines, 11, 15, 37, 42, 43, 44, 45, 46, 87, 92

H

hazards, 31, 42
height, 11, 12, 49, 57, 76
House, 4, 30, 36, 54, 82, 102, 103
House of Representatives, 4, 36, 54, 82
Housing and Urban Development, 49
human, 10, 17, 31, 37, 102
hurricane barriers, vii, 3, 4, 22, 24, 25, 26, 56
Hurricane Katrina, 25, 33, 36, 47
hurricanes, 32
hydroelectric power, 102
hydropower, vii, 4, 10, 32, 52, 53, 54, 55, 59, 66, 67, 71, 76, 78, 79, 82, 86, 89, 92, 101, 103

I

identification, 67
improvements, 19, 22, 24, 25, 32, 88
infrastructure, vii, 2, 3, 4, 5, 6, 7, 8, 9, 10, 11, 13, 14, 15, 16, 17, 20, 22, 23, 24, 25, 26, 28, 32, 54, 56, 67, 82, 86, 91
Inland Waterways Trust Fund (IWTF), 51, 76, 78
inspections, 36, 49, 59, 91, 102
integration, 22
internal controls, 64
irrigation, 76, 82, 99, 102
issues, 9, 17, 21, 31, 32, 34, 37, 55, 86, 93

L

land acquisition, 40
laws, 34, 37, 52, 55, 83, 86
lead, 20, 28, 34, 36, 37, 42, 47, 63, 67, 77, 86, 94
leadership, 9, 91

legislation, 9, 10, 29, 38, 44, 46, 47, 53, 63, 84, 87, 95, 99
Levee Safety, v, 23, 24, 31, 33, 36, 37, 38, 40, 41, 42, 43, 46, 49, 50
levees, vii, 2, 3, 4, 21, 22, 23, 24, 26, 31, 33, 35, 36, 37, 38, 39, 40, 41, 42, 43, 45, 46, 47, 48, 49, 50, 56
life cycle, 8, 15, 17
limestone, 66
litigation, 92
local community, 8
local government, 8, 9, 36, 42, 43, 76, 82, 86

M

majority, 36, 48, 67, 76
management, vii, 2, 4, 5, 13, 15, 21, 22, 25, 26, 32, 40, 41, 42, 43, 45, 46, 49, 50, 56, 83, 87, 91, 92, 94, 102, 103
materials, 39, 42, 56
measurements, 11, 12, 13, 17
melting, 83
military, 6, 7, 56, 76, 85, 86, 101
mission, 14, 26, 86, 97, 102
Mississippi River, 76
Missouri, 14, 30, 87, 97, 101, 103
models, 13, 18, 20, 22, 62, 86, 96
modifications, 23, 24, 32, 61, 64, 86, 91
moisture, 12, 14, 20, 103
Montana, 86

N

National Academy of Sciences, 9
National Aeronautics and Space Administration, 29
National Data Buoy Center (NDBC), 1, 12
National Environmental Policy Act of 1969 (NEPA), 1, 30, 92
National Oceanic and Atmospheric Administration (NOAA), 1, 4, 5, 6, 12, 21, 22, 28, 29, 30, 32, 49, 79, 82, 87, 98
National Research Council, 4, 82, 100
national security, 30
National Security Council, 15, 30
National Water Center (NWC), 1, 12, 13
National Weather Service (NWS), 1, 6, 12, 13, 18, 20, 22, 30, 87
natural disasters, 32, 40, 50
natural gas, 66
natural hazards, 31
Natural Resources Conservation Service (NRCS), 1, 5, 6, 11, 12, 49
navigation, vii, 4, 7, 10, 27, 32, 76, 82, 101, 103
NEPA, 1, 30

O

Office of Management and Budget, 10, 14, 89
officials, 2, 5, 6, 9, 13, 14, 17, 18, 19, 20, 21, 22, 23, 24, 25, 27, 28, 29, 30, 32, 34, 36, 38, 42, 44, 46, 47, 52, 55, 58, 60, 61, 62, 63, 64, 65, 66, 67, 69, 78, 80, 81, 83, 84, 89, 90, 91, 92, 93, 94, 96, 97, 99, 101, 102
Oklahoma, 73, 101
operations, vii, 2, 5, 6, 7, 10, 13, 14, 15, 16, 18, 19, 26, 27, 31, 32, 39, 52, 55, 59, 70, 77, 79, 80, 81, 82, 83, 84, 85, 86, 87, 89, 90, 93, 94, 95, 96, 97, 99, 100
operations and maintenance (O&M), 59, 70, 79, 85, 89, 93
opportunities, 14, 17, 31
outreach, 40, 59
oversight, 7, 23, 56, 86

P

peer review, 9, 17, 29, 30, 91
permeability, 76

policy, 7, 14, 15, 30, 34, 38, 52, 53, 56, 60, 61, 63, 64, 65, 67, 71, 72, 86
population, 25, 36
population density, 25
power generation, 102
precipitation, 4, 12, 13, 21, 30, 38, 81, 82, 84, 95, 97
preparation, 2, 5, 45, 46, 83
preservation, 32
President, 7, 16, 30, 56, 86, 89, 93, 98, 102
prevention, 32
private sector, 41
probability, 23, 42, 57, 96, 102
production costs, 60
project, 3, 5, 8, 9, 10, 15, 16, 17, 18, 19, 25, 26, 27, 28, 29, 30, 31, 32, 49, 50, 55, 58, 59, 60, 61, 62, 67, 68, 69, 70, 71, 75, 76, 77, 79, 80, 81, 82, 83, 84, 85, 86, 87, 88, 90, 91, 92, 93, 94, 95, 96, 97, 99, 101, 102
protection, 32, 37, 38, 39
public education, 42, 43, 45, 46

Q

quality control, 9, 91

R

rainfall, 17, 18, 21, 97, 101
Reclamation (Bureau of Reclamation), 6, 29, 31, 49, 51, 64, 78, 96, 98, 102
recommendations, iv, 9, 35, 37, 38, 40, 41, 44, 45, 46, 52, 72, 80, 100
recovery, 40, 53, 68
recreation, 10, 32, 76, 82, 86, 101, 102, 103
redundancy, 8
Reform, 2, 5, 34, 35, 37, 38, 41, 44, 46, 47, 48, 80, 83, 100
regulations, 15, 18, 31, 41, 52, 55, 63, 67, 80, 84, 87, 91, 93
rehabilitation, 36, 42, 43, 45, 46, 49, 78

relief, 34, 36, 47
repair, 25, 32, 49, 50, 52, 53, 54, 55, 56, 58, 59, 61, 62, 63, 65, 67, 68, 69, 70, 71, 72, 77
requirements, 3, 9, 14, 19, 22, 24, 31, 38, 41, 43, 47, 52, 55, 61, 71, 83, 92, 93, 100
resilience, 14, 16, 17, 93, 98, 102
resistance, 68
resolution, 9, 29, 67, 98
resources, vii, 2, 3, 4, 5, 6, 7, 8, 11, 13, 15, 16, 17, 21, 22, 25, 28, 30, 34, 46, 47, 48, 50, 56, 82, 83, 84, 86, 87, 91, 93, 94, 99, 102
response, 3, 5, 13, 14, 19, 22, 23, 40, 68, 72, 82, 91, 92
restoration, 7, 25, 32, 103
risk, 3, 10, 14, 15, 17, 22, 23, 24, 25, 26, 27, 31, 32, 33, 36, 38, 40, 41, 43, 47, 49, 50, 53, 54, 57, 58, 62, 64, 71, 76, 77, 88, 91, 92, 96, 99, 101, 102, 103
risk assessment, 3, 22, 23, 24, 25, 26, 31, 41, 49, 102
risk management, 15, 25, 40, 49, 50, 92
rules, 96
runoff, 13, 17, 20, 28, 82, 83, 92, 97

S

SACE, 102
safety, vii, 2, 5, 6, 20, 23, 30, 31, 33, 34, 35, 36, 37, 38, 41, 42, 43, 44, 45, 46, 47, 48, 49, 52, 53, 54, 55, 56, 57, 58, 59, 60, 61, 62, 63, 64, 65, 66, 67, 68, 69, 70, 71, 72, 76, 77, 78, 89, 91, 94, 96
salinity, 31
saltwater, 32
science, 9, 16, 19, 26, 28, 29, 82, 84, 91, 92, 97
sea level, 3, 5, 16, 22, 26, 27
Secretary of Defense, 48, 72, 100
Secretary of Homeland Security, 48
sediment, 17, 87, 94, 98, 103
sedimentation, 94, 97, 98

Index

seismic data, 53, 61, 63, 64
Senate, 4, 35, 54, 82
services, 13, 29, 56, 76, 101
shoreline, 25, 30
Snow Telemetry (SNOTEL), 1, 11
Soil Climate Analysis Network (SCAN), 1, 12
solution, 58, 61, 77
Southeastern Power Administration (SEPA), 51, 65, 66, 67, 70, 73, 77, 78
species, 92
specifications, 11, 58
stability, 57
stabilization, 68
stakeholders, 8, 10, 15, 16, 20, 22, 36, 37, 40, 47, 48, 68, 80, 81, 84, 87, 92, 97, 98
state, vii, 3, 4, 8, 9, 12, 21, 22, 29, 31, 35, 40, 41, 42, 43, 45, 46, 47, 52, 53, 55, 56, 59, 61, 63, 64, 65, 66, 67, 68, 70, 71, 72, 76, 77, 78, 80, 86, 91, 98, 102
statutory authority, 8, 52, 64
storage, 10, 18, 76, 87, 91, 98, 101, 102
storms, 4, 13, 32, 82
structure, 38, 69, 85, 86

T

Task Force, 14, 49, 50
technical assistance, 37, 42, 45, 46
technical comments, 28, 49
technical support, 77
techniques, 9, 21, 78, 96, 97, 102, 103
technology, 19, 88, 97
temperature, 12, 95
threats, 5, 16, 98
training, 36, 42, 45, 46, 86
transactions, 94, 99
transmission, 60
transport, 103
Treasury, 60, 68
treatment, 71
Trust Fund, 51, 76, 78

U

U.S. Army Corps of Engineers (Corps), v, vii, 1, 2, 3, 4, 5, 6, 7, 8, 9, 10, 11, 12, 13, 14, 15, 16, 17, 18, 19, 20, 21, 22, 23, 24, 25, 26, 27, 28, 29, 30, 31, 32, 33, 34, 35, 36, 37, 38, 40, 41, 42, 43, 44, 45, 46, 47, 48, 49, 50, 51, 52, 53, 54, 55, 56, 57, 58, 59, 60, 61, 62, 63, 64, 65, 66, 67, 68, 69, 70, 71, 72, 75, 76, 77, 78, 79, 80, 81, 82, 83, 84, 85, 86, 87, 88, 89, 90, 91, 92, 93, 94, 95, 96, 97, 98, 99, 100, 101, 102, 103
U.S. Department of Agriculture, 29
U.S. Department of the Interior, 64
U.S. Geological Survey (USGS), 1, 2, 5, 6, 11, 12, 13, 18, 20, 22, 29, 49, 79, 82, 87, 98, 101
U.S. Treasury, 60, 68
United States, v, 1, 4, 11, 12, 17, 29, 30, 32, 33, 35, 36, 43, 50, 51, 54, 76, 79, 82, 95, 100
universe, 81, 94, 99
universities, 21
updating, 2, 14, 15, 16, 17, 19, 102
urban areas, 39

V

vapor, 103
variables, 98
vegetation, 39
vulnerability, 3, 6, 22, 25, 26, 27, 28, 32, 62, 97, 98

W

Washington, 7, 21, 29, 30, 31, 32, 49, 56, 77, 85, 98, 100, 101, 102, 103
water, vii, 2, 3, 4, 5, 6, 7, 8, 10, 11, 12, 13, 14, 15, 16, 17, 18, 19, 20, 22, 26, 27, 28, 29, 30, 31, 32, 38, 50, 52, 54, 55, 56, 57, 59, 61, 63, 68, 69, 70, 71, 72, 76, 77, 79, 80, 81, 82, 83, 84, 85,

Index

86, 87, 88, 89, 90, 91, 92, 93, 94, 96, 98, 99, 100, 101, 102, 103
Water Control, v, 18, 73, 79, 86, 87, 89, 90, 91, 100, 101
water resources, vii, 2, 3, 4, 5, 6, 7, 8, 11, 13, 15, 16, 17, 21, 22, 28, 30, 50, 56, 82, 83, 86, 87, 93, 94, 99
Water Resources Development Act of 1986 (WRDA), 1, 9, 10, 29, 30, 51, 60, 61, 63, 78
water shortages, 82

water supply, vii, 4, 11, 32, 55, 59, 68, 69, 70, 71, 72, 76, 79, 81, 82, 88, 97, 99, 101, 102, 103
water vapor, 103
watershed, 7, 19, 20, 28, 56, 85
Weather Events, v, 1, 16, 22, 26, 49
weather patterns, 5, 82
wetlands, 30
White House, 30, 102, 103
wholesale, 60, 73
wildlife, 32, 82, 101, 102
wildlife conservation, 82